River Variability and Complexity

Rivers differ among themselves and through time. An individual river can vary significantly downstream, changing its dimensions and pattern dramatically over a short distance. If hydrology and hydraulics were the primary controls on the morphology and behavior of large rivers, we would expect long reaches of rivers to maintain characteristic and relatively uniform morphologies. In fact, this is not the case – the variability of large rivers indicates that other important factors are involved.

River Variability and Complexity presents a new approach to the understanding of river variability. It provides examples of river variability and explains the reasons for them, including fluvial response to human activities. Understanding the mechanisms of variability is important for geomorphologists, geologists, river engineers and sedimentologists as they attempt to interpret ancient fluvial deposits or anticipate river behavior at different locations and through time. This book provides an excellent background for graduates, researchers and professionals.

Stanley A. Schumm is an internationally recognized geomorphologist, whose primary experience has been the investigation and analysis of fluvial systems. During his career Dr Schumm has worked as a geologist with the United States Geological Survey and taught at Colorado State University for thirty years. He has served on numerous professional, technical and government advisory committees in the US and around the world. He is the recipient of many prizes and awards for his scientific contributions and papers, including the Kirk Bryan Award of the Geological Society of America for his book *The Fluvial System* and in 1997 the GSA's Distinguished Career Award. He has published two other books with Cambridge. These are, *To Interpret the Earth: Ten Ways to Be Wrong* (1991) and *Active Tectonics and Alluvial Rivers* (2000, co-authored with J. F. Dumont and J. M. Holbrook).

River Variability and Complexity

Stanley A. Schumm
Mussetter Engineering, Inc., USA

CAMBRIDGE
UNIVERSITY PRESS

CAMBRIDGE UNIVERSITY PRESS
Cambridge, New York, Melbourne, Madrid, Cape Town, Singapore, São Paulo

Cambridge University Press
The Edinburgh Building, Cambridge, CB2 2RU, UK

Published in the United States of America by Cambridge University Press, New York

www.cambridge.org
Information on this title: www.cambridge.org/9780521846714

First published 2005

Printed in the United Kingdom at the University Press, Cambridge

A catalog record for this book is available from the British Library

Library of Congress Cataloguing in Publication data
Schumm, Stanley Alfred, 1927–
River variability and complexity / Stanley A. Schumm.
    p.   cm.
Includes bibliographical references (p.).
ISBN 0-521-84671-4
1. Rivers.   2. Geomorphology.   3. Sedimentation and deposition.   I. Title.
GB1203.2.S362 2005
551.48′3 – dc22   2004051272

ISBN-13 978-0-521-84671-4 hardback
ISBN-10 0-521-84671-4 hardback

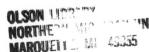

To my grandchildren
Katie and William Richardson
Emily and Jackson Stewart

May they enjoy their work as much as I have mine.

Contents

Preface

The origin of this book was a project involving the geomorphic character of the lower Mississippi River and water-rights litigation between the Forest Service and the State of Colorado. The Mississippi River project revealed to me that large alluvial rivers can vary greatly in morphology downstream, although hydrologic conditions are not greatly different. This suggests that river-control works and activities such as dredging will influence a river differently depending upon channel variability and the diverse character of reaches. The water rights litigation confirmed that generalizations about rivers, such as hydraulic-geometry relations have limits depending upon scale.

It must be recognized that rivers differ among themselves, and through time, and one river can vary significantly in a downstream direction. If the morphology and behavior of large alluvial rivers are determined primarily by hydrology and hydraulics, long reaches of alluvial rivers should maintain a characteristic and relatively uniform morphology. In fact, this is not the case, and the variability of large alluvial rivers is an indication that hydraulics and hydrology are not always the dominant controls. Therefore, the purpose of this book is to present to the fluvial community examples of river variability and the reasons for them. The recognition that marked changes from one type of river or river pattern to another can occur is important for geomorphologists, river engineers, and stratigraphers.

The normal variability of a large river is not the topic here. Mark Twain beautifully describes this in one of his books *The Gilded Age* (Twain 1873, pp. 41–43) as:

> Sometimes the boat fought the mid-stream current, with a verdant world on either hand, and remote from both; sometimes she closed in under a point, where the dead water and the helping eddies were, and shaved the bank so closely that the decks were swept by the jungle of over-hanging willows and littered with a spoil of leaves; departing from these "points" she regularly crossed the river every five miles, avoiding the "bight" of the great bends and thus escaping the strong current; sometimes she went out and skirted a high "bluff" sand-bar in the middle of the stream, and occasionally followed it up a little too far and touched upon the shoal water at its head – and then the intelligent craft refused to run herself aground, but "smelt" the bar, and straightway the foamy streak that streamed away from her bows vanished, a great foamless wave rolled forward and passed her under way, and in this instant she leaned far over on her side, shied from the bar and fled square away from the danger like a frightened thing – and the pilot was lucky if he managed to "straighten her up" before she drove her nose into the opposite bank; sometimes she approached a solid wall of tall trees as if she meant to break through it, but all of a sudden a little crack would open just enough to admit her, and away she would go plowing through the "chute" with just barely room

enough between the island on one side and the main land on the other; in this sluggish water she seemed to go like a racehorse, sometimes she found shoal water, going out at the head of those "chutes" or crossing the river, and then a deck-hand stood on the bow and hove the lead, while the boat slowed down and moved cautiously.

This description of the Mississippi River reveals considerable variability, but it is the normal variability of a great alluvial river. This discussion of rivers deals with and explains the anomalies of river form and behavior.

A good example of what is intended is provided by Simpson and Smith (2001) in their description of the variability of the Milk River in Alberta, Canada and Montana, USA. The meandering Milk River in Canada suddenly becomes braided in Montana. The difference is attributed to bank material variability and channel widening in the braided reach, which is accompanied by reduced stream power.

There is probably little that is entirely new in this book, but the organization of controls on river morphology may be useful (Figure 1.2). Often all that the modern researcher can do is to quantify the observations of earlier workers. For example, while working on the Great Plains, I recognized that bed and bank sediments appeared to control the width–depth ratio and sinuosity of these rivers. Imagine my chagrin to discover that an English engineer, W. Jessop had stated this in 1782: "Where rivers run through a country where the soil is pure clay, loam or any thing of light and homogenous quality, they are always very deep, and in general narrow; on the contrary, where they run through a soil that has in its composition a considerable mixture of sand, gravel, or other hard matter, they always become wide or shallow . . ." (Petts, 1995, p. 8). There appears to be little that is new. Nevertheless, the objective here is to provide geomorphologists, engineers, sedimentologists, and stratigraphers with descriptions and explanations of the downstream variability of rivers.

Acknowledgments

Once again, I thank Saundra Powell for her skill and kindness in producing an acceptable copy of this work for the publishers. It is remarkable that she appears to enjoy processing what I write. Also sincere thanks to Mussetter Engineering, Inc. for considerable assistance in preparation of figures and providing a place to work. Many thanks to Bonnie Vail for many kindnesses and to Matt Iman for drafting many of the figures.

Colleagues whose help was essential for work in difficult locations are acknowledged with gratitude: Louis Flam, Pakistan, Mohen jo Daro; Victor Galay, Egypt, River Nile; and Jack Mabbutt, Central Australia, Finke River.

Finally, an apology to all of my colleagues whose work I have not cited. A book could be written for each topic listed in Figure 1.2, which is beyond me.

Part I

Background

Chapter 1

Introduction

Upon having some astronomical phenomena explained to him, Alfonso X, King of Castile and Leon (1252–84) exclaimed,

> If the Lord Almighty had consulted me before embarking upon creation, I should have recommended something simpler
>
> (Mackay, 1991)

River engineers and geomorphologists might well have a similar opinion especially when it is recognized how variable a river can be through time and from reach to reach. However, when Leopold and Maddock published US Geological Survey Professional Paper 252 it was a landmark occasion. Geologists and geomorphologists suddenly became aware of order in rivers, although engineers with their regime equations had anticipated these hydraulic geometry relations. The hydraulic geometry relations of width, depth, and velocity were immediately of value in prediction of river characteristics. However, some of us neglected to recognize how variable the relations were and how significant was the scatter about the regression lines. This should have warned us that, yes, in a general sense channel width increased downstream as the 0.5 power of discharge, but a prediction of what the width was around the next bend could be in gross error, and, therefore recognizing this variability could be of considerable practical significance.

River characteristics vary sometimes little and sometimes greatly. Reaches are singular because of the numerous variables acting that prevent a single variable, discharge, from dominating river morphology and behavior. The question to be answered is why is one reach of a river connected to a different type of reach? That is, why can reaches be so different? For example, why does a straight river become meandering and a meandering river braid or anabranch? An understanding is critical to the practical application of river data.

Recently, books dealing with this fluvial variability have been edited (Gregory, 1977; Schumm and Winkley, 1994; Gurnell and Petts, 1995; Miller and Gupta, 1999). Most of the literature dealing with river variability and change has involved what have been referred to as alluvial rivers or alluvial adjustable rivers, and these have been grouped into braided, meandering, and straight. However, more recently considerable attention has been devoted to the study of steep mountain streams and the effects of bedrock (Tinkler and Wohl, 1998).

A modern alluvial river is one that flows on and in sediments transported by the river during the present hydrologic regime, but it is associated with an older sediment complex at depth. Alluvial rivers have always played an important role in human affairs. All of the early great civilizations rose on the banks of large alluvial rivers such as the Nile, Indus, Yellow, Tigris, and Euphrates. River engineering began in those early times to minimize the effects of floods and channel changes. Today, engineers face the same problems, and they have been successful in developing flood control, navigation, and channel stabilization programs but often at great cost and with the need to continually maintain and repair structures and channels.

In order to manage alluvial rivers, an understanding of their complexity in space and through time is necessary. They differ in three ways:

1. there is a spectrum of river types that is dependent upon hydrology, sediment loads, and geologic history (in other words, rivers differ among themselves);
2. rivers change naturally through time as a result of climate and hydrologic change;
3. there can be considerable variability of channel morphology along any one river, as a result of geologic and geomorphic controls (Schumm and Winkley, 1994).

Information on these differences, especially the last two, will aid in predicting future river behavior and their response to human activities.

An important consideration in predicting future river behavior and response is the sensitivity of the channel. That is, how readily will it respond to change or how close is it to undergoing a change without an external influence? For example, individual meanders frequently develop progressively to an unstable form, and a chute or neck cutoff results, which leads to local and short-term channel adjustments. The cutting off of numerous meanders along the Mississippi River caused dramatic changes, as a result of steepening of gradient, which led to serious bank erosion and scour (Winkley, 1977).

Because of this complexity the stratigrapher-sedimentologist, who must interpret ancient valley-fill and alluvial plain deposits, faces a great challenge. For example, many fluvial successions will display characteristics of more than one type of river. This is not "sedimentological anarchy," as suggested by Walker (1990), but it is a recognition of the complexity and variability of fluvial systems in space and time (Miall, 1996, p. 202).

If the sedimentologist-stratigrapher is concerned with the vertical third dimension of an alluvial deposit, the river engineer and geomorphologist is essentially concerned with the two-dimensional surface of the valley fill.

The term fluvial is from the Latin word *fluvius*, a river. When carried to its broadest interpretation a fluvial system not only involves stream channels but also entire drainage networks. The size of fluvial systems ranges from that of the vast Mississippi, Missouri, and Ohio river system to small badland watersheds of a few square meters. The time periods that are of interest to the student of the fluvial system can range from a few minutes of present-day activity, to channel changes of the past century, to the geologic time periods required for the development of the billion-year-old gold-bearing paleo-channels of the Witwatersrand conglomerate and even older and spectacular channels and drainage networks on Mars. Therefore, the range of temporal and spatial dimensions of the fluvial system is very large.

In order to simplify discussion of the complex assemblage of landforms that comprise a fluvial system, it can be divided into three zones (Figure 1.1). Zone 1 is the drainage basin, watershed, or sediment-source area. This is the area from which water and sediment are derived. It is primarily a zone of sediment production, although sediment storage does occur there in important ways. Zone 2 is the transfer zone, where, for a stable channel, input of sediment can equal output. Zone 3 is the sediment sink or area of deposition (delta, alluvial fan). These three subdivisions of the fluvial system may appear artificial because obviously sediments are stored, eroded, and transported in all the zones; nevertheless, within each zone one process is dominant.

Each zone, as defined above, is an open system. Each has its own set of morphological attributes, which can be related to water discharge and sediment movement. For example, the divides, slopes, floodplains, and channels of Zone 1 form a morphological system. In addition, the energy- and materials-flow form another, that of a cascading system. Components of the morphological system (channel width, depth, drainage density) can be related statistically to the cascading system (water and sediment movement, shear forces, etc.) to produce a fluvial process-response system.

The fluvial system can be considered at different scales and in greater or lesser detail depending upon the objective of the observer. For example, a large segment, the dendritic drainage pattern is a component of obvious interest to the geologist and geomorphologist (Figure 1.1a). At a finer scale there is the river reach of Figure 1.1b, which is of interest to those who are concerned with what the channel pattern reveals about river history and behavior, and to engineers who are charged with maintaining navigation and preventing channel erosion. A single meander can be the dominant feature of interest (Figure 1.1c), which is studied by geomorphologists and hydraulic engineers for information that it provides on flow hydraulics, sediment transport, and rate of bend shift. Within the channel itself is a sand bar (Figure 1.1c), the composition of which is of concern to the sedimentologist, as are the bed forms (ripples and dunes) on the surface of the bar (Figure 1.1d) and the details of their sedimentary structure (Figure 1.1e). This, of course, is composed of the individual

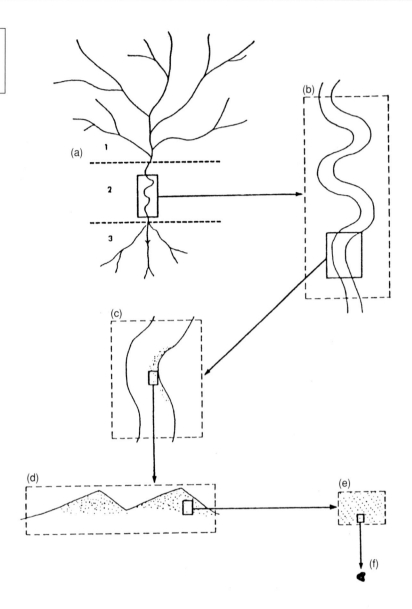

Figure 1.1 Idealized sketch showing the components of the fluvial system. See text for discussion (after Schumm, 1977).

grains of sediment (Figure 1.1f) which can provide information on sediment sources, sediment loads, and the feasibility of mining the sediment for construction purposes or for placer deposits.

As the above demonstrates, a variety of components of the fluvial system can be investigated at many scales, but no component can be totally isolated because there is an interaction of hydrology, hydraulics, geology, and geomorphology at all scales. This emphasizes that the entire fluvial system cannot be ignored, even when only a small part of it is under investigation. Furthermore, it is important to realize that although the fluvial system is a physical system, it follows an evolutionary development, and it changes through time. Therefore, there are a great variety of rivers in space, and they change

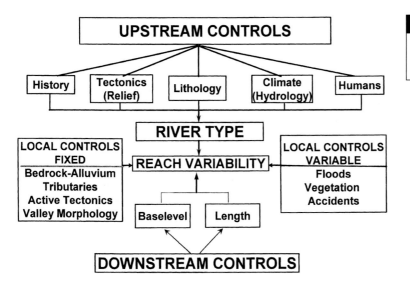

Figure 1.2 Chart showing various controls of river morphology and behavior (after Piégay and Schumm, 2003).

Figure 1.3 Index map showing location of many rivers in the mid-continent USA that are discussed in this volume.

through time in response to upstream (Figure 1.1a, Zone 1) and downstream (Zone 3) controls.

Figure 1.2 provides an outline of this book as it lists the variables that determine river type and reach variability and Figure 1.3 shows the location of many US rivers that are discussed later. Three types of control determine what happens at a reach: (1) general controls determine the type of river (braided, meandering, anastomosing, wandering); and (2) the downstream control of base-level and length modify the effects of the general controls. At the reach scale, however, (3) local controls can dominate. They can be fixed in the sense that their position changes little or they can be variable changing in location and time.

As noted above, it is important to recognize how these variables control river behavior and valley-fill sedimentology. This awareness may aid the engineer in his plans and the sedimentologist in his interpretation of fluvial sediments.

Human activity takes place everywhere (Figure 1.2), but these impacts will not be considered, except as an upstream control (Figure 1.2), as they are usually obvious. For example, riprap, dikes, diversions, etc. can be constructed anywhere, and they are fully discussed in the engineering literature (Peterson, 1986) and by experts in the field of human impacts on rivers (Brizga and Finlayson, 1999; Wohl, 2000a; Anthony *et al.*, 2001). Nevertheless, human involvement with rivers for better or worse is considered in Part 6 (Chapters 18, 19, and 20).

Chapter 2

Types of rivers

Before considering the variability of a single river, it is necessary to consider the different types of rivers that exist (Table 2.1). Once a topic is sufficiently comprehended, it appears logical to develop a classification of its components. A classification can provide a direction for future research, and there have been many attempts to classify rivers (e.g., Schumm, 1963; Mollard, 1973; Kellerhals *et al.*, 1976; Brice, 1981; Mosley, 1987; Rosgen, 1994; Thorne, 1997; Vandenberghe, 2001). Indeed, Goodwin (1999) thinks that there is an atavistic compulsion to classify, and indeed, an individual's survival may depend on an ability to distinguish different river types (deep versus shallow).

Depending upon the perspective of the investigator, a classification of rivers will depend upon the variable of most significance. For example, the classic braided, meandering and straight tripart division of rivers (Leopold and Wolman, 1957) is based upon pattern with boundaries among the three patterns based upon discharge and gradient. Brice (1982, 1983) added an anabranched or anastomosing channel pattern (Figure 2.1) to the triad and distinguished between two types of meandering channels (Table 2.1). The passive equiwidth meandering channel is very stable as compared to the wide-bend point-bar meandering channel (Figure 2.2). This is a very important practical distinction between active and passive meandering channels (Thorne, 1997, p. 188). A highly sinuous equiwidth channel gives the impression of great activity whereas, in fact, it can be relatively stable (Figure 2.3). Brice also indicates how width, gradient, and sinuosity, as well as type of sediment load and bank stability varies with pattern (Figure 2.2).

Based upon examination of sand-bed streams of the Great Plains (Kansas, Nebraska, Wyoming, Colorado), USA, and the Murrumbidgee River, Australia, Schumm (1968) proposed a three-part division of rivers based upon type of sediment load and channel stability (Table 2.2). The bed sediment in these channels did not vary significantly; therefore, grain size was not related to channel morphology, but type of load (suspended, mixed, or bed-material) was.

There are five basic bed-load channel patterns (Figure 2.4) that have been recognized during experimental studies. These five basic

Table 2.1	Channel types

Regime (Graded) channels
Patterns
    straight
    meandering (passive/active)
    wandering
    braided
    anastomosing (can be any of above patterns)
Hydrology
    ephemeral
    intermittent
    perennial
    interrupted

Non-regime channels
Bedrock
    confined
    constrained
Unstable
    aggrading (transport limited)
    degrading (supply limited)
    avulsing

bed-load channel patterns can be extended to mixed-load and suspended-load channels to produce 13 patterns (Figure 2.4). Patterns 1–5 are bed-load channel patterns as defined in Table 2.2. Patterns 6–10 are mixed-load channel patterns, and patterns 11–13 are suspended-load channel patterns. The patterns change with increasing valley slope, stream power, and sediment load for each channel type.

As compared to the bed-load channel pattern, which have high width–depth ratios (Table 2.2), the five-mixed load patterns (Figure 2.4b) are relatively narrower and deeper, and there is greater bank stability. The higher degree of bank stability permits the maintenance of narrow-deep straight channels (Pattern 6), and alternate bars stabilize because of the finer sediments, to form slightly sinuous channels (Pattern 7). Pattern 8 is a truly meandering channel, wide on the bends, relatively narrow at the crossings, and subject to chute cutoffs. Pattern 9 maintains the sinuosity of a meandering channel, but with the larger sediment transport the presence of bars gives it a wandering sinuous-braided appearance. Pattern 10 is an island-braided channel that is relatively more stable than that of bedload channel 5.

Suspended-load channels (Figure 2.4c) are narrow and deep. Suspended-load Pattern 11 is a straight, narrow, deep channel. With only small quantities of bed load, this type of channel will have the highest sinuosity of all (Patterns 12 and 13) but as noted above, the channel can be very stable.

Transitional patterns such as Brice's braided point bar channel (Figure 2.2) can be termed wandering. A wandering river as described

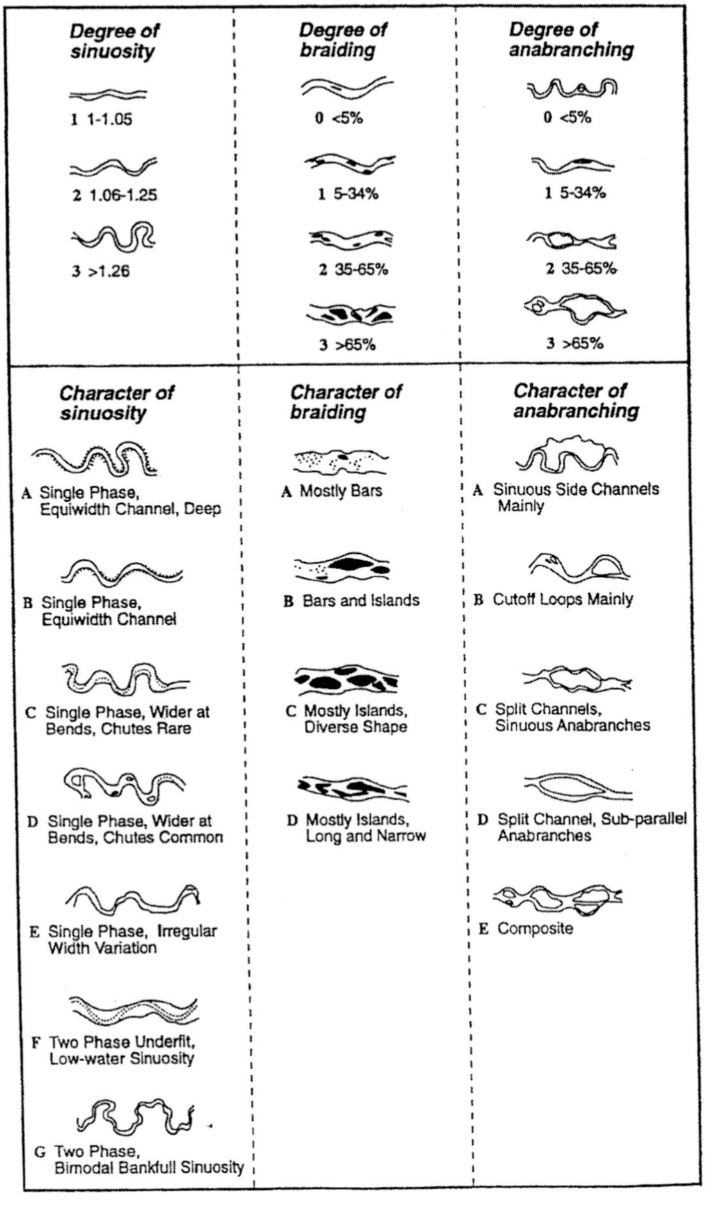

Figure 2.1 Channel pattern classification (after Brice, 1975).

by Church (1983, pp. 179–80) "exhibits an irregular pattern of channel instability. Although a single dominant channel is everywhere evident, the river consists of a sequence of braided/anastomosed reaches connected by relatively stable single-thread reaches." Desloges and Church (1987, p. 99) expand the description to stress that a wandering river "exhibits an irregularly sinuous channel, sometimes split about channel islands and in some places braided . . ." The wandering river appears to be a transitional pattern between active meandering and braiding, and it may not be in regime.

The channels of the previous discussion are all alluvial rivers as defined before. That is, the channel is not confined by bedrock or

Figure 2.2 Lateral stability of different channels with changes of morphology and type of sediment load (after Brice, 1982).

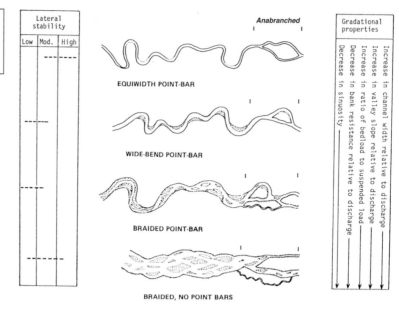

Figure 2.3 Plot showing that at a given width, equiwidth meandering channels are more stable than wide-bend meandering channels (after Brice, 1982). Note that six equiwidth channels show essentially no movement.

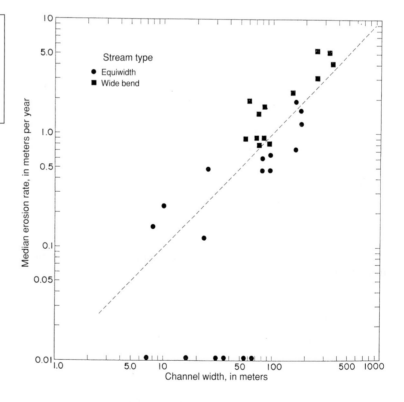

terraces, but it is flanked by a floodplain. This type of channel is what British engineers who were designing and constructing canals in Egypt and India termed regime channels (Table 2.1). When the bed and banks of a channel "scour and fill, changing depth, gradient and width until a state of balance is attained . . . the channel is said to be

Mode of sediment transport and type of channel	Bedload (percentage of total load)	Channel stability		
		Stable (graded stream)	Depositing (excess load)	Eroding (deficiency of load)
Suspended load	<3	Stable suspended-load channel. Width/depth ratio <10; sinuosity usually >2.0; gradient, relatively gentle	Depositing suspended load channel. Major deposition on banks cause narrowing of channel; initial streambed deposition minor	Eroding suspended-load channel. Streambed erosion predominant; initial channel widening minor
Mixed load	3–11	Stable mixed-load channel. Width/depth ratio >10, <40; sinuosity usually <2.0, >1.3; gradient, moderate	Depositing mixed-load channel. Initial major deposition on banks followed by streambed deposition	Eroding mixed-load channel. Initial streambed erosion followed by channel widening
Bed load	>11	Stable bed-load channel. Width/depth ratio >40; sinuosity usually <1.3; gradient, relatively steep	Depositing bed-load channel. Streambed deposition and island formation	Eroding bed-load channel. Little streambed erosion; channel widening predominant

Table 2.2 Classification of alluvial channels

(From Schumm 1977.)

in regime" (Lindley, 1919). Regime channels (Lacey, 1930, Lane, 1937), are analogous to the geomorphologist's graded rivers (Mackin, 1948), although Mackin's discussion concentrated on changes of gradient through time.

Brice shows (Figure 2.2) that for each of his river types there exists a corresponding multiple channel river (anabranching or anastomosing) and Nanson and Knighton (1996) have identified two types of straight anastomosing channels (Figure 2.5).

It must be stressed that the preceding classification applies to adjustable alluvial rivers, with sediment loads primarily of sand, silt, and clay, and they would be identified as regime channels by Montgomery and Buffington (1997) who have considered the full range of channels from high mountain bedrock channels to those described previously (see also, Rutherfurd, 1999). They start at the drainage divide (Figure 2.6) and move down through bedrock and colluvial depressions or chutes to the point where one can recognize alluvial channels. They identify five distinct reach morphologies (Figure 2.7):

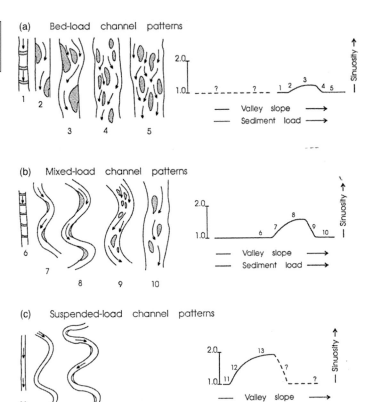

Figure 2.4 River patterns based upon type of sediment load and gradient (after Schumm, 1977).

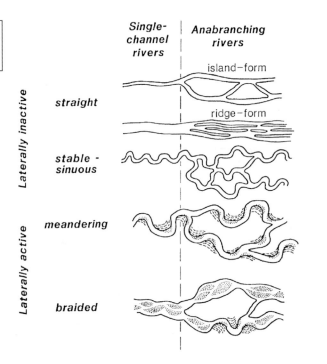

Figure 2.5 Types of anabranching channels (after Nanson and Knighton, 1996).

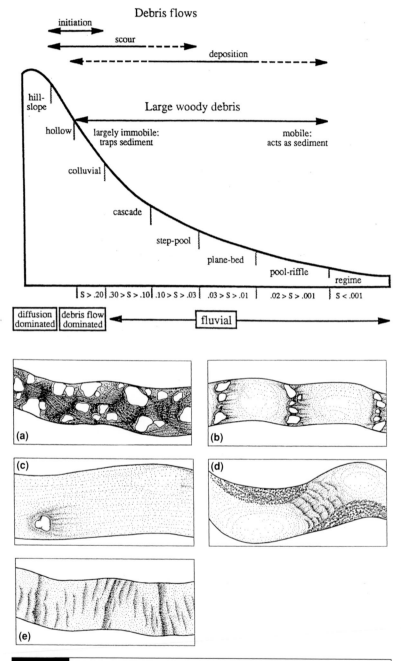

Figure 2.6 Idealized long profile from hillslopes and unchanneled hollows downslope through the channel network showing the general distribution of channel types and controls on channel processes in mountain drainage basins (from Montgomery and Buffington, 1997).

Figure 2.7 Schematic planform illustration of channel morphologies at low flow: (a) cascade channel (Figure 2.6) showing nearly continuous, highly turbulent flow around cobbles and boulders; (b) step-pool channel showing sequential highly turbulent flow over steps and more tranquil flow through intervening pools; (c) plane-bed channel showing single boulder protruding through otherwise uniform flow; (d) pool-riffle channel showing exposed bars, highly turbulent flow through riffles, and more tranquil flow through pools; and (e) dune-ripple channel showing dune and ripple forms as viewed through the flow (from Montgomery and Buffington, 1997).

Table 2.3	External influences on mountain streams	
Geologic	*Organic*
Bedrock (presence)	Vegetation
Bedrock (variability)	presence
Joints (spacing)	type
Faults	log jams
	Beavers
Geomorphic	Human
Alluvial fans	roads
Mass movement	mines
debris flow	timbering
boulder fall	
Glacial	*Hydraulic*
outwash	Armoring
moraine	Backwater
Terraces	bars
	benches
	Boulder steps

cascade, step pool, plane bed, pool riffle, and dune ripple (regime). Most of these reaches will be confined by valley walls and terraces in contrast to the regime channels, and indeed, there are a number of external influences on these channels (Table 2.3).

Channel size and location with regard to bedrock controls can dominate mountain river morphology and behavior. Montgomery and Buffington's (1997) classification of mountain streams illustrates this well with the smallest streams being dominated by debris flows and large woody debris, which are factors that are absent or have little impact along most large alluvial rivers. Mountain streams form a continuum from total bedrock control to regime. In fact, mountain streams can be divided simply into three groups: confined, constrained, and regime.

A confined channel (Table 2.1) is totally controlled by bedrock or old alluvium in the banks and bed. Tinkler and Wohl (1998) characterized a bedrock channel as one with 50 percent bedrock exposed in the bed and banks. It is very clear that such channels are not self-adjusting, and the knowledge and relations developed for regime channels of the lowlands cannot be transferred to confined and constrained channels. The confined channel is static in the short-term although, during geologic time, the river can deepen its valley and widen the valley floor. Although hydraulic geometry-type relations may be developed between valley (channel) width and discharge, this does not signify that the channel is in any sense in regime.

The constrained channel (Table 2.1) is subjected to all of the previously listed controls with colluvium, bedrock, and terrace alluvium determining the dimensions and morphologic character of the channel. The degree of control will decrease from left to right on Figure 2.6, but again, relations developed for the regime channel cannot be extrapolated to the constrained channel.

The hydrologic character of a river is another important control on channel morphology (Table 2.1). Consider how different ephemeral flow, intermittent flow, and perennial flow channels can be. Also, flashy discharge will have a major impact different from that of a channel with relatively uniform flow. An exotic river flowing from areas of high to low precipitation will reverse the normal hydraulic geometry relations (see Figures 7.1, 7.2, 7.3). Ross (1923, p. 36) describes interrupted streams as having permanent flow over short reaches throughout the year while most of the river is dry.

Classifications of rivers concentrate on regime or "stable" channels that although changing position through time, nevertheless maintain a standard morphology. Unstable channels that are incising, aggrading, or avulsing are treated separately or not at all. Hence, any single classification will be inadequate, and separate classifications are probably required for braided, meandering, and straight channels (Figure 2.4). Non-regime channels can be aggrading, degrading, or avulsing for a variety of reasons (see Chapter 3) and these changes can be different for different rivers (Table 2.2).

Although classifications are initially useful for clarity of communication and as an index of the numerous types of channels that exist (Rosgen, 1994), it is really the characteristics of an individual channel that is important. From a practical perspective, the geomorphologist's measurements of sinuosity, width–depth ratio, gradient, dimensions (width, depth), and sediment type (bed and bank) when combined with the engineer's measurements of discharge, flow velocity, and stream power, provide the information necessary for understanding of a river and the knowledge required for prediction of future change. When quantitative information about a river is available, classifications are of less value in the design of stable channels and the prediction of channel change. Clearly, the variability of an alluvial river will have many causes.

To summarize, rivers can be identified as regime and non-regime (Table 2.1). Regime channels display a variety of patterns and flow characteristics. Non-regime channels are controlled by bedrock or older alluvium or they are unstable.

It should be recognized that most channel classifications are based upon channel pattern. However, each pattern is associated with a particular shape and gradient. Therefore, recognition of a pattern should provide some information on other morphologic characteristics (Figures 2.2, 2.3, 2.4, Table 2.2). Finally, it should be noted that each river that is in regime has a floodplain, the stratigraphy of which reflects river type (Melton, 1936; Nanson and Croke, 1992).

Chapter 3

Non-regime channels

Except for the mountain streams discussed in Chapter 2, all of the other channels were alluvial and in regime. Therefore, it is appropriate here to spend some time considering channels that are responding to altered conditions by eroding, depositing, avulsing, and changing pattern – i.e., non-regime channels.

Rivers respond to altered conditions, therefore variability through time is important because a river or a reach may be out of character for a period as it adjusts. For example, Table 2.2 suggests how different types of channels respond to change. A list of 16 responses of channels to change, and the four major variables that influence them are summarized in Table 3.1. Time (history) is included with discharge (increase or decrease), sediment load (increase or decrease), and base level change (up or down), because channels change naturally through time, and time is an index of energy expended or work done. The changes are grouped according to the results of the change (erosion, deposition, pattern change, and metamorphosis). In Table 3.1, the changes that will be affected by the passage of time or by a change of discharge, sediment load or base level are indicated by an X. A brief discussion of each of the responses follows, but two – incision and avulsion – will be considered in detail because of their potential for causing serious problems. Based upon experience, a reader may disagree with the channel responses of Table 3.1, and this is expected because of the great variability and complexity of rivers and reaches.

Erosion

Degradation (incision) is the lowering of a streambed by erosion. It is not local scour, rather it is a major adjustment of a river to external controls. The adjustment takes place over long reaches to form an incised channel. The deepening of the channel may also cause bank erosion and widening. Degradation will reduce the width–depth ratio of a channel, and when the incision is great and the reduction of the

Table 3.1 | Variables affecting channels. An X indicates a channel response

Channel response	Time	Variables					
		Discharge		Sediment load		Base level	
		+	−	+	−	up	down
Erosion							
Degradation (incision)		X			X		X
Nickpoint formation and migration		X			X		X
Bank erosion	X	X		X		X	X
Deposition							
Aggradation			X	X		X	
Back and downfilling			X	X		X	
Berming			X	X			
Pattern change							
Meander growth and shift	X	X					
Island and bar formation and shift	X	X		X			
Cutoffs	X	X		X		X	X
Avulsion	X	X		X		X	
Metamorphosis							
Straight to meandering		X		X			X
Straight to braided			X	X		X	X
Braided to meandering		X			X		X
Braided to straight			X		X		X
Meandering to straight		X		X		X	X
Meandering to braided			X	X		X	

(From Schumm *et al.*, 1983.)

width–depth ratio is large, the channel is considered to be incised (Schumm *et al.*, 1984; Darby and Simon, 1999).

Channel incision has been and is a major concern because it disrupts transportation, destroys agricultural land, threatens adjacent structures, drastically alters environmental conditions, and produces sediment that causes further problems downstream (Cooke and Reeves, 1976; Graf, 1983; Schumm *et al.*, 1984; Simon, 1994). Therefore, the causes of channel incision have been a topic of great interest because a better understanding of the phenomenon could lead to prevention. We do know a great deal about the causes of incision, and it is a prime example of equifinality or convergence (Schumm, 1991) because there are many causes of channel incision, but they all produce the same general result.

The causes of channel incision can be grouped into six categories (Table 3.2) that at least partly reflect the different time scales at which formative processes operate. Geologic and geomorphic causes may require many years to develop a response, whereas climatic and hydrologic variability, animal grazing, and human impacts can occur during shorter periods of time. In addition, the climatic and hydrologic

Table 3.2 | Causes of incision

A Geologic
 1 uplift (Burnett and Schumm, 1983; Ouchi, 1985)
 2 subsidence (Ouchi, 1985)
 3 faulting (Keller and Pinter, 1996; Mayer, 1985)
 4 lateral tilt (Reid, 1992)

B Geomorphic
 1 stream capture (Shepherd, 1979; Galay, 1983)
 2 baselevel lowering (Begin et al., 1980, 1981)
 3 meander cutoffs (Winkley, 1994; Love, 1992)
 4 avulsion (Kesel and Yodis, 1992; Schumm et al., 1996)
 5 lateral channel shift (Lamarche, 1966; Galay, 1983)
 6 sediment storage (increased gradient) (Patton and Schumm, 1965; Trimble, 1974; Meade et al., 1990; Vandaele et al., 1996)
 7 mass movement (Martinson, 1986; Rodolfo, 1989; Ohmori, 1992)
 8 ground-water sapping (Schumm and Phillips, 1986; Higgins, 1990; Jones, 1997)

C Climatic
 1 drier (Knox, 1983; Graf, 1983; Hall, 1990)
 2 wetter (Knox, 1983; Graf, 1983; Bull, 1991)
 3 increased intensity (Knox, 1983; Balling and Wells, 1990)

D Hydrologic
 1 increased discharge (Burkard and Kostachuk, 1995)
 2 increased peak discharge (Macklin et al., 1992)
 3 decreased sediment load (Lane, 1955; James, 1991)

E Animals
 1 grazing (Alford, 1982; Graf, 1983; Prosser and Slade, 1994)
 2 tracking (Trimble and Mendel, 1995)

F Humans
 1 dam construction (Galay, 1983; Williams and Wolman, 1984)
 2 sediment diversion (Galay, 1983)
 3 flow diversion (Maddock, 1960; Bray and Kellerhals, 1979)
 4 urbanization (Morisawa and Laflure, 1979; Booth, 1991)
 5 dam removal, failure (Galay, 1983)
 6 lowering lake, reservoir levels (Born and Ritter, 1970; Galay, 1983)
 7 meander cutoff (Lagasse, 1986; Winkley, 1994)
 8 underground mining (Goudie, 1982)
 9 ground-water and petroleum withdrawal; hydrocompaction (Lofgren, 1969; Goudie, 1982; Prokopovich, 1983)
10 gravel mining (Lagasse, 1986; Harvey and Schumm, 1987)
11 dredging (Lagasse, 1986)
12 roads, trails, ditches (Burkard and Kostachuk, 1995; Gellis, 1996)
13 channelization (Schumm et al., 1984; Simon, 1994)
14 constriction of flow (Petersen, 1986)
15 deforestation (Macklin et al., 1992; Burkard and Kostaschuk, 1995)
16 fire (Laird and Harvey, 1986; Heede et al., 1988)

(From Schumm, 1999.)

causes can be closely related. For example, a wetter climate (C2) will increase discharge (D1), and increased rainfall intensity (C3) will increase peak discharge (D2). Also, there is feedback from human and animal activities to hydrologic controls. For example, human activities (F: 1, 3, 4, 5, 13, 15, and 16) will modify water discharge and sediment loads.

Geologic causes (Table 3.2)

Uplift, subsidence, and faulting all modify the slope of a valley floor and channel gradient. For example, channel incision should occur on the downstream steeper part of an uplift and at the upstream steeper side of subsidence and where a stream crosses from the upthrown to the downdropped portions of a fault. Lateral tilt of a valley floor can cause avulsion and the development of a new channel downdip, and lateral fault displacement can also cause stream incision and gullies.

The magnitude of the resulting incision in each case depends on the amount of deformation and the ability of a channel to adjust to the altered slope perhaps by an increase of sinuosity (Schumm, 1993). It has even been suggested that fractures caused by an earthquake were instrumental in causing incision of the San Pedro River, Arizona, USA during subsequent floods (Hereford, 1993).

Geomorphic causes (Table 3.2)

Geomorphic causes of incision for the most part involve an increase of gradient. For example, stream capture, base-level lowering, meander cutoffs, avulsion, cliff retreat, sediment storage, and lateral shift of a main channel can cause a local steepening of channel gradient and incision. It has been demonstrated, for example, that sediment deposition and storage in valleys and on alluvial fans eventually leads to steeper gradients and the formation of discontinuous gullies and alluvial-fan trenches (Schumm et al., 1984, 1987), when a threshold of slope stability or stream power is exceeded (Begin and Schumm, 1979; Bull, 1979).

Other processes, such as mass movement, can produce incision. Mudflows are capable of erosion and channel enlargement (Rodolfo, 1989), and when a landslide delivers large amounts of sediment to a valley floor, this deposit will eventually be incised by the existing stream. In addition, where groundwater emerges from a sloping surface, a channel can be produced that extends back toward a drainage divide or to the source of groundwater (Schumm et al., 1995).

Climatic and hydrologic causes (Table 3.2)

Climatic and hydrologic causes of incision are closely related, and perhaps they should not be separated in Table 3.2. Nevertheless, it is possible to think of a climate change that alters vegetational type and cover during a period of years in contrast to short-term climatic fluctuations that increase discharge and flood peaks without significantly affecting vegetation. A climatic change to the drier can reduce vegetational cover and produce higher sediment loads and higher peak discharges. A change to a wetter climate can increase vegetative cover, which reduces sediment loads and increases mean annual runoff. These climatic changes produce hydrologic changes that cause incision. Climatic fluctuations produce hydrologic responses such as major floods, periods of increased discharge, and changes of sediment loads.

Animal causes (Table 3.2)

It has long been argued that overgrazing has produced the great arroyos of southwestern USA. Whether or not this is true, weakening of vegetational cover by grazing can produce channel incision, as can the concentration of flow in animal trails. Tracking of animals also decreases infiltration rates, which in turn increases runoff both on hillslopes and in the trails, which leads to gully development (Blackburn et al., 1982; Trimble, 1988; Trimble and Mendel, 1995).

Human causes (Table 3.2)

Human activity of a variety of types is known to cause channel incision. These causes can be grouped into four effects: decreased sediment loads, increased annual discharge and peak discharge, flow concentration, and increased channel gradient. Decreased sediment loads can be caused by dam construction, urbanization, diversion of sediment into another channel such as a canal, and gravel mining and dredging. Mean discharge and peak discharge can be increased by flow diversion, urbanization, dam removal, and deforestation. Flow can be concentrated with the effect of increasing stream power by gravel mining and dredging, roads, trails, ditches, channelization, and by flow constriction by dikes. Gradient can be increased by dam removal, lowering of water levels in lakes and reservoirs, meander cutoffs, mining, withdrawal of fluids and hydrocompaction, gravel mining, dredging, and channelization.

Controls of incision

Once incision has commenced, it is unlikely that erosion will cease naturally until the channel has progressed through the several stages of the incised channel evolution model (Figure 3.1). Incised channels after initial incision, follow an evolutionary sequence that results in relative stability after a period of years, and once channel incision has begun, it is unlikely to cease unless humans control the channel. Figure 3.1 is based upon the evolution of channelized streams in Mississippi and Figure 3.2 is based upon similar studies in New Mexico. Differences in incised channel morphology depend upon the stage of evolution and upon the type of sediment forming the channel (Table 2.1). In addition, the type of sediment in the alluvial valley fill can significantly affect the process of channel incision and adjustment. For example, it was determined that channels incising into cohesive sediments (silts and clays) deepen rapidly, whereas channels incising into sandy sediments widen rapidly and deepen much less (Table 2.1). In addition, deposition in the different channels also follows a different pattern (Table 2.1). Hence, the natural

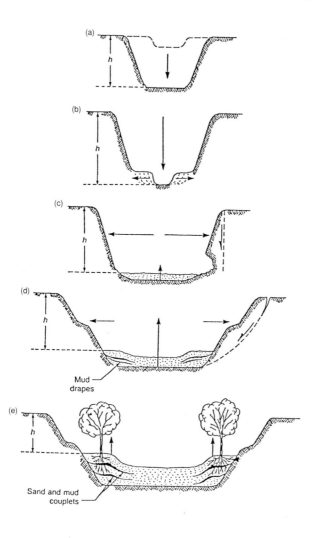

Mud
drapes

Sand and mud
couplets

Figure 3.1 Evolution of an incised channel from initial incision (a, b) and widening (c, d) to aggradation (d, e) and eventual stability (e) (modified after Schumm *et al.*, 1984). The dashed cross-section (a) represents the pre-incision channel.

control of incised-channel change is primarily sedimentologic, which influences the rate, type, and magnitude of incision.

For cost-effective control of incising channels, timing is most important and, therefore, it is an important variable in any scheme to control an incised channel and to reduce sediment loads. Figure 3.3 shows, in general, how incised channels change with time and how sediment yield follows this trend. In a drainage basin that contains incising channels, sediment production increases as the size of incising channels increases (Figure 3.3, Times a–c). However, at Time c maximum, growth of the incised channels has occurred, and they begin to stabilize between Times d and e. By understanding this cycle of channel incision from initial stability (Time a) to renewed stability (Time f) it is possible to select times when channel control practices will be most effective. For example, when incision is just commencing (Times a and b) or when channels are almost stabilized (Times e and f) control measures will be most successful in preventing incision and finally stabilizing the channels. However, at Times c and d, control will be difficult and expensive.

Figure 3.2 Hypothetical sequence for geomorphic evolution of a large arroyo based on the Rio Puerco (Figure 1.3). Bar represents channel width (after Elliott et al., 1999).

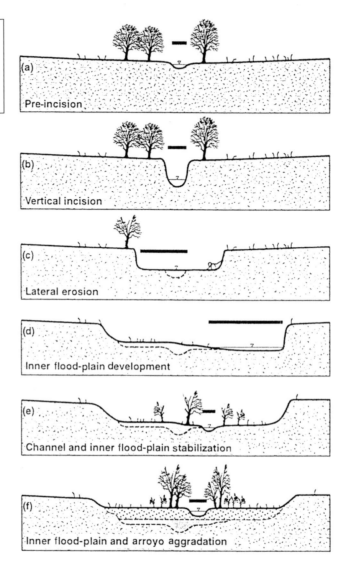

(a) Pre-incision

(b) Vertical incision

(c) Lateral erosion

(d) Inner flood-plain development

(e) Channel and inner flood-plain stabilization

(f) Inner flood-plain and arroyo aggradation

Figure 3.3 Hypothetical changes in sediment production and active-channel width over time, as it evolves from stage a to stage f (Figure 3.1). Dashed lines indicate effect of gully-control structures at various times during channel evolution (from Schumm, 1999).

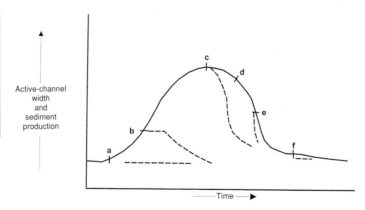

Active-channel width and sediment production

Time

Although Figure 3.3 was developed independently of Figures 3.1 and 3.2, they can be related. For example, the pre-incision channel of Figure 3.1(a) is associated with Time a (Figure 3.3), whereas the newly incised channel of Figure 3.1(a) occurs between Times a and b (Figure 3.3). The still deepening channel of Figure 3.1(b) occurs between Times b and c of Figure 3.3. The widening and commencement of sediment storage of Figure 3.1(c) occurs between Times d and e of Figure 3.3. The relatively mature channel of Figure 3.1(e) develops between Times d and e, and is complete at Time f (Figure 3.3). This sequence can take about 40 to 50 years in channelized streams of the southeastern (Figure 3.1) USA and over 100 years in the arroyos of the southwestern (Figure 3.2) USA.

In summary, there are many causes of channel incision (Table 3.1), therefore, each incised channel must be evaluated separately. However, in many cases the incision process destroys evidence for its cause. An understanding of the cause of incised channels is important for their prevention, but it is probably less so for their control following incision. In many cases, more than one cause combines to induce incision. For example, a large flood (D2) may trigger incision, but it is the buildup of sediment in the valley (B7) that sets the stage for incision, and the impact of animals or humans may also be significant.

Although all of the causes listed on Table 3.1 can cause channel incision, they may not. Depending upon the magnitude of the impact and the sensitivity of the channel or valley floor, incision does not always result. Therefore, it is important to recognize that in a given region, channel behavior may not be in phase. That is, some channels may be incised, whereas others are aggrading or are relatively stable. Observation of some incised channels should not cause an investigator to ignore other apparently stable valley floors. Finally, timing is important regarding control of incised channels, and the results of human impacts can vary greatly depending upon the stage of incised-channel evolution (Figure 3.3).

Other responses

The other erosional responses listed on Table 3.1 are obvious. For example, nickpoint migration is the upstream shift of an inflection in the longitudinal profile of the stream (Figure 3.4). This break in the smooth curve of the stream gradient results from rejuvenation and incision of the stream. A nickpoint in alluvium moves upstream, especially during floods. Above the profile break the river is stable; below the break there is erosion. As the nickpoint migrates past a point, a dramatic change in channel morphology and stability occurs (Schumm et al., 1987). Nickpoints are of two types: first is a sharp break in profile which forms an in-channel scarp called a headcut (Figure 3.4a), and second is a steeper reach of the channel, a nick-zone over which the elevation change is distributed (Figure 3.4b). It is important to recognize that through time a stable reach of river may become suddenly very unstable as a result of passage of a nickpoint.

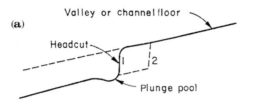

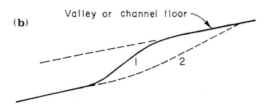

Figure 3.4 Types of nickpoints: (a) a steep headcut; and (b) a sloping nickzone. Numbers show present (1) and future (2) positions of channel floor.

Bank erosion (Table 3.1) is the removal of bank materials by either a grain-by-grain removal or by mass failure. The effect of bank erosion is a shift in the bank line of the river and the introduction of additional sediment into the channel. Erosion of both banks widens the channel, and it may lead to aggradation. During and following major floods a river transporting a high bedload and therefore, one with a readily erodible floodplain will deepen moderately, but bank erosion will cause the channel to widen greatly. Therefore, width–depth ratio will be large. Comparison of relatively small semiarid streams in the western USA shows how differently they respond (Table 2.2) depending upon type of sediment load and floodplain characteristics. For example, the Gila (Burkham, 1972), Cimarron (Schumm and Lichty, 1963), and Salado Rivers (Bryan, 1927) behaved as bed load channels (Table 2.1) with minor incision, but major widening in response to large floods.

Deposition

Aggradation as considered in Table 3.1, is simply the raising of a stream bed by deposition, but, aggradation may continue to the extent that new changes occur. For example, it may cause avulsion, meanders to cut off, and channel pattern change. In addition, aggradation may lead to bank erosion as flow paths are changed by bar formation, and decreased channel capacity will increase flooding.

The many causes of incision listed on Table 3.2 can also cause aggradation. For example, the geologic causes can create a gentler slope as well as a steeper one. The result, therefore, is aggradation. Some of the geomorphic controls when reversed can cause aggradation. For example, base-level rise, meander growth, avulsion to a flatter area and blockage of channels by mass movement all lead to deposition.

Climatic and hydrologic changes, when they are the reverse of that of Table 3.2, can also cause aggradation. A number of human activities such as dam construction, sediment and flow diversion, raised reservoir and lake levels, as well as groundwater and petroleum withdrawal leading to subsidence all can cause aggradation. Of course, all of the causes of incision greatly increase downstream sediment loads which result in downstream aggradation. Aggrading channels, unless transporting only silts and clays, will be braided. Often a scouring reach of channel will be converted downstream to braiding, as the eroded sediment is deposited.

Backfilling (Table 3.1) is deposition or channel filling from downstream to upstream. That is, the channel is partly or entirely blocked and deposition begins at this point and then proceeds upstream (Schumm, 1977, p. 150). Backfilling differs from aggradation as defined earlier because it starts at one location in the channel and then is propagated upstream. In contrast, downfilling occurs when deposition progresses in a downstream direction and it is the reverse of backfilling. Both backfilling and downfilling are types of aggradation that influence long reaches of a channel, and they can affect a reach of river from either the upstream or downstream direction after it has been stable for a long time. Consequences of backfilling and downfilling are similar to those of general aggradation. Increased flooding will result as the channel fills. Deposition may also cause deflection of the thalweg of the channel to the extent that bank erosion will become important, as the channel attempts to find its way around the more recently deposited sediment. Finally, berming (Table 3.1) refers to the deposition of sediments on the sides of the channel, and it is the opposite of bank erosion. Berming will reduce the area of the channel and cause increased flood stages. Berming reduces channel capacity, but the narrowing of the channel may cause degradation and scour.

Pattern change

A variety of channel changes are listed under pattern change (Table 3.1). For example, meander growth and shift involves a change in the dimensions and position of a meander (Figure 3.5). Meander amplitude and width increase, as a meander enlarges. Meander shift involves the displacement of the meander in a downstream direction (Figure 3.5e). Usually the meander both grows and shifts downstream, although some parts of the bend can actually shift upstream. Meander growth and shift not only cause bank erosion at the crest and on the downstream side of the limbs of a meander, but it also changes the flow alignment. Further, increased meander amplitude results in a local reduction of gradient with possible aggradation in the bend. Meander growth and shift will be of greatest significance where discharge is great, bank sediments are weak, and bank vegetation is negligible due to aridity or to agricultural practices.

Island and bar formation and shift (Table 3.1) are within-channel phenomena (Figure 3.5a and b). Unlike meander shift or meander

Figure 3.5 Typical channel changes. Dashed lines show future channel position.

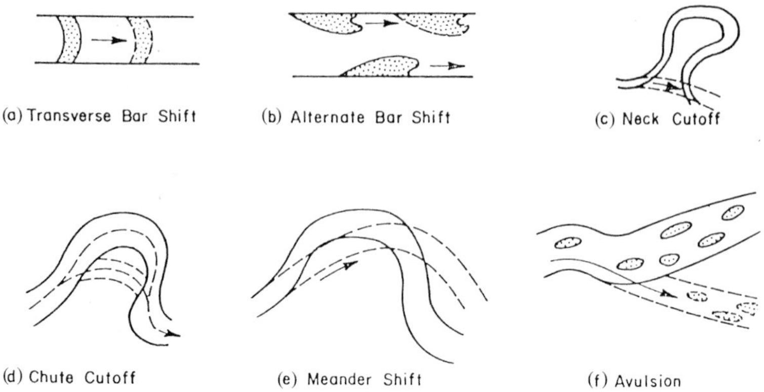

(a) Transverse Bar Shift (b) Alternate Bar Shift (c) Neck Cutoff

(d) Chute Cutoff (e) Meander Shift (f) Avulsion

Figure 3.6 Island change. Arrows show direction of flow. Solid lines are original locations of islands; dashed lines show changes. (a) Island shifts up or downstream; (b) island shifts laterally; (c) island divided by channel; (d) small islands coalesce and island joins floodplain; and (e) islands increase or diminish in size (from Popov, 1962).

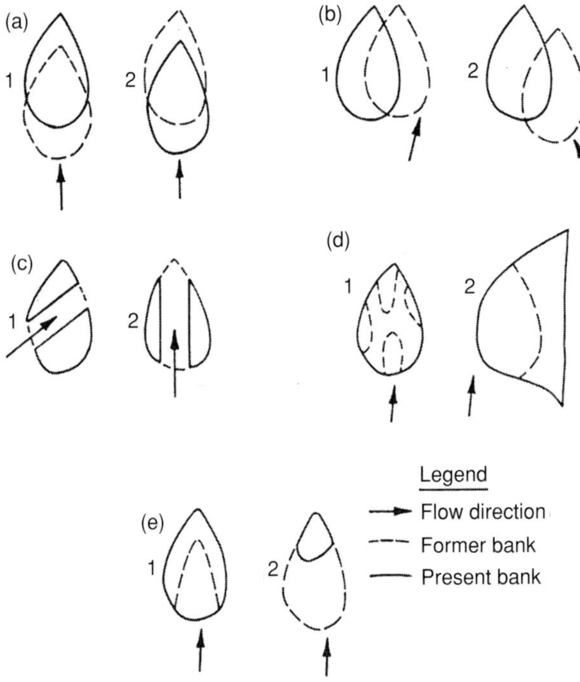

Legend
— Flow direction
--- Former bank
— Present bank

cutoffs, which involve the entire channel pattern, bars and islands can evolve within the channel, and the bankline pattern itself may remain unchanged.

Popov (1962) has classified the types of island changes that he observed occurring in the River Ob in the former Soviet Union. He found that there were five ways islands change (Figure 3.6). A sixth and seventh could be added; that is, the formation of an island and the complete destruction of an island, but Figure 3.6 does convey the important concept that bars and islands may be ephemeral as well as dynamic features of a channel. The result of bar and island formation in a channel is to deflect the flow and perhaps to increase erosion of the banks of the channel. This erosion will enlarge the channel and islands may form as a result of reduced water stage.

Avulsion

Avulsion (Table 3.1) is the abrupt change of the course of a river (Figure 3.5f). A channel is abandoned and a new one formed as the water and sediment take a new course across the floodplain. A meander cutoff (Table 3.1) is one type of avulsion because it is a relatively rapid change in the course of the river during a short period of time. This drastically reduces the length of the stream in that reach and significantly steepens its gradient. The neck cutoff has the greatest effects (Figure 3.5c) on the channel. Another type of cutoff is the chute cutoff (Figure 3.5d), which forms by cutting across a portion of the point bar.

The consequences of cutoffs of both types is that the river is steepened abruptly at the point of the cutoff. This can lead to scour at that location and a propagation of the scour in an upstream direction. The results are similar to those described for degradation and nickpoint migration. In the downstream direction, the gradient of the channel is not changed below the site of the cutoff, and therefore, the increased sediment load caused by upstream scour will usually be deposited below the cutoff, forming a large bar, or it may trigger additional downstream cutoffs. However, Brice (1982) and Stevens (1994) have demonstrated that some cutoffs do not induce significant channel change because of the resistance of floodplain sediments. Avulsion can also involve a major change of channel character both above and below the point of avulsion. If, through avulsion, the river takes a shorter course to the sea, it will have a steeper gradient, and erosion above the point of avulsion is likely unless a bedrock control prevents upstream degradation.

Avulsion is a major threat to human activities and large sums of money are spent to prevent avulsion of the Mississippi River down the shorter Atchafalaya River (Figure 1.3) route to the Gulf of Mexico (Schumm, 1977, p. 303). The Yellow River is perched 10 m above the north China Plain (Figure 3.7). It is constrained by rock levees which prevent avulsion (Zhou and Pan, 1994). The lower Indus River in Sindh, Pakistan is also perched on a convexity (Jorgensen *et al.*, 1993) and it has a history of major avulsions. The Brahmaputra River avulsed about 100 km to the west from east of Dakka to west of it (Figure 5.3), during the late nineteenth century (Schumm *et al.*, 2000, p. 116) perhaps as a result of a large earthquake. Obviously avulsions of this magnitude have had a great impact on river navigation and other types of transportation; the loss of life could be great, and the economy of a region could be devastated.

In braided streams, the term avulsion is sometimes used to describe the shift of the main thread of current to the other side of a mid-channel bar, but the term should be restricted to the complete shift of the entire channel (Figure 3.5f).

The length of channel reach that is affected by an avulsion can vary from a few tens of meters (Jones and Harper, 1998), to hundreds of kilometers (Fisk, 1944). The rapidity with which discharge shifts completely from an old to a new channel varies from as little as one day (Ning, 1990) to a decade (Jones and Harper, 1998) or much longer. Avulsion may result in the rapid shift of flow to a single new channel (Brizga and Finlayson, 1999) or in a gradual process of crevasse splay formation, wetland infilling, and channel coalescence (Smith *et al.*, 1989, Ethridge *et al.*, 1999).

Avulsion commonly results when an event (usually a flood) of sufficient magnitude occurs along a reach of a river that is at or near an avulsion threshold. As demonstrated for geomorphic systems in general, the closer a river is to the threshold, the smaller is the event needed to trigger the avulsion. This is why avulsions are not always triggered by the largest floods on a given river (Brizga and Finlayson, 1999; Ethridge *et al.*, 1999).

The underlying causes of avulsions (i.e., those processes or events that move a river toward an avulsion threshold) can be organized into four groups (Table 3.3). The first two groups involve an increase in the ratio, S_a/S_e. S_a is the slope of the potential avulsion course, and S_e is the gradient of the existing channel. In Group 1, the increase in S_a/S_e is due to a decrease in S_e. The decrease in S_e will reduce the ability of the channel to carry water and sediment. In Group 2, the increase in S_a/S_e is caused by an increase in S_a and does not necessarily result in channel blockage. In the third group, the ability of the channel to carry water and sediment decreases for reasons unrelated to slope changes. The fourth group is composed of causes which do not fall into the first three categories.

In Group 1 (Table 3.3), S_e can decrease as a result of four processes, leading to in-channel deposition and avulsion. A pattern change (1a) leading to increased sinuosity will cause avulsion. Increased sinuosity leads to slope reduction, resulting in bedload deposition where the slope decreased, clogging of the channel, and finally, avulsion. A good example of this is the Ovens River in southeastern Australia, which avulsed from a sinuous channel into a steeper straight channel. The greater velocity and power in the straight channel causes incision

Table 3.3 | Causes of avulsion

Processes and events that create instability and lead toward avulsion		Ability of channel to carry sediment and discharge
Group 1 – Avulsion from increase in ratio, S_a/S_e^*, due to decrease in S_e	a. Sinuosity increase (reduced gradient)	Decrease
	b. Delta growth (lengthening of channel)	Decrease
	c. Baselevel fall (resulting in decreased slope)	Decrease
	d. Tectonic uplift (resulting in decreased slope)	Decrease
Group 2 – Avulsion from increase in ratio, S_a/S_e, due to increase in S_a	a. Natural levee/alluvial ridge growth	No change
	b. Alluvial fan and delta growth (convexity)	No change
	c. Tectonism (resulting in lateral tilting)	No change
Group 3 – Avulsion with no change in ratio, S_a/S_e	a. Hydrologic change in flood peak discharge	Decrease
	b. Increased sediment load	Decrease
	c. Vegetative encroachment	Decrease
	d. Log jams	Decrease
	e. Ice jams	Decrease
Group 4 – Other avulsions	a. Animal trails	No change
	b. Capture (diversion into adjacent drainage)	No change

(After Jones and Schumm, 1999.)

$^*S_a$ is the slope of the potential avulsion course, S_e is the slope of the existing channel.

and bank erosion that eventually forms a sinuous channel, which is again susceptible to avulsion (Schumm *et al.*, 1996).

Figure 3.8 is a map of the Ovens and King Rivers complex of avulsed channels. There are four generations of channels. The youngest is Deep Creek which is a young incised channel that has and is capturing the flow from the Ovens River. A cross-section of the valley (Figure 3.9) shows the alluvial ridges of the almost abandoned Tea Garden Creek, the abandoned Tarrawingee channel, the still active Ovens River, and the newly formed Deep Creek.

This example shows how reaches of the Ovens River have been abandoned by avulsion and given different names. The sinuosity of the older channels is about 2.0, whereas the sinuosity of Deep Creek is 1.2. As Deep Creek captures more of the Ovens Creek water and sediment, it will widen, become shallower, and develop a more sinuous course. Eventually it will develop a high sinuosity, and avulsion will eventually cause its abandonment.

The extension of deltas (Table 3.3:1b) at constant sea level also decreases the gradient of the existing channel. Where distributaries

Figure 3.8 Map of anastomosing channels of Ovens and King Rivers, Australia. Numbers indicate relative ages of channels and reaches of channels from youngest (1) to oldest (4) (from Schumm et al., 1996).

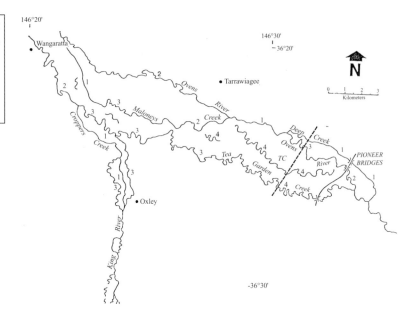

Figure 3.9 Cross-section across Ovens River anastomosing system of channels. Dashed line on Figure 3.8 is location of cross-section (from Schumm et al., 1996).

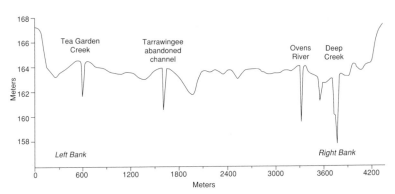

are relatively straight and laterally stable (e.g., Mississippi River delta), delta lobe growth lengthens the channel and reduces slope, creating a threshold condition. Flooding may then trigger avulsion down a shorter and steeper slope to the sea. This type of avulsion appears to be common on fluvial dominated deltas where delta lobes extend seaward. Examples include the Mississippi (Fisk, 1944), the Po (Nelson, 1970), and Yellow Rivers (Ning, 1990; Van Gelder et al., 1994). In many deltas, net subsidence away from the active delta front may simultaneously increase gradient.

A base-level fall (Table 3.3:1c) that exposes a seafloor of lower slope than that of the existing river channel will produce a reduced gradient where the channel meets the seafloor surface and may result in channel deposition and avulsion. In a similar manner, uplift or faulting (Table 3.3:1d) across a river course can decrease slope and cause a river to avulse around the obstruction. Avulsions in the second group (Table 3.3) result from increased lateral slope (S_a) away from the existing channel. A major cause is the development of an alluvial ridge

(Table 3.3:2a) where the channel and levees aggrade more rapidly than the surrounding floodplain (Fisk, 1944). The channel bottom (thalweg) of a river on an alluvial ridge is commonly above the level of the floodplain (Schumann, 1989; Brizga and Finlayson, 1999; Bryant et al., 1995; Heller and Paola, 1996), although elevation of the thalweg above the adjacent floodplain is not necessary in order to reach the avulsive threshold (Schumm et al., 1996). Alluvial ridge development (Table 3.3:2a) and sinuosity increase (Table 3.3:1a) commonly combine to create a perched channel and a gradient decrease, as described for the Ovens and King Rivers in Australia (Figure 3.8) and Red Creek in Wyoming (Schumann, 1989). In this case, decrease of S_e and increase of S_a combine to increase the slope ratio S_a/S_e.

An upward convex shape of a delta or alluvial fan (Table 3.3:2b) creates a setting (similar to natural levee and alluvial ridge development) in which the channel becomes perched on the highest point of the deposit, then avulses down the steepest course.

Increase of S_a due to lateral tilting (Table 3.3:2c) of a valley floor from tectonism or other causes can produce repeated avulsion in one direction, as has been described for the South Fork Madison River (Leeder and Alexander, 1987), St. Francis River (Boyd and Schumm, 1995), and Owens River (Reid, 1992).

The third group (Table 3.3) involves a reduction in the capacity of a channel to convey all of the water and sediment delivered to it. In this group, sudden channel blockage – e.g. a channel-blocking mass movement – can trigger an avulsion, whereas more gradual channel blockage moves the river toward a threshold condition in which a periodic flood may serve as the trigger. An increase of flood peaks (Table 3.3:3a) can cause increased overbank flooding and avulsion if the existing channel cannot accommodate the increased flow.

In-channel deposition (Table 3.3:3b) from a variety of causes other than slope change can produce channel blockage and result in avulsion. Schumann (1989) describes how bank failure along Red Creek, an ephemeral stream in Wyoming, leads to the formation of within-channel benches of fine sediment that reduce channel width and cause overtopping and avulsion during floods. Along Red Creek, the divergence point is at the downstream end of a meander bend where super elevation of the water surface promotes overbank flooding. Sediment delivered by tributaries, mass failure, or a general increase in sediment load from upstream can all cause in-channel deposition, clogging, and avulsion. Sand dune migration into a channel may result in avulsion in eolian settings (Jones and Blakey, 1997), particularly where ephemeral streams are present.

Vegetation encroachment (Table 3.3:3c) may decrease channel capacity. McCarthy et al. (1992) showed how growth of channel-lining vegetation restricted channel flow causing in-channel sedimentation and eventual avulsion on the Okavango fan, Botswana. This type of biological activity may be important in some settings.

Log jams (Table 3.3:3d) can cause avulsion by blocking the channel and forcing overbank flow. One massive jam, the great Red River Raft in Louisiana, was nearly 160 km long and was not cleared to permit navigation of the river until 1876. For a fascinating discussion of the problems involved in the removal of this impressive feature see McCall (1984).

Ice-jam induced flooding (Table 3.3:3e) is a very common occurrence and may be an important contributing factor to avulsion in some climates. Ethridge *et al.* (1999) document an ice-jam induced avulsion on the Niobrara River in Nebraska. A major ice jam in 1928 may have triggered an avulsion of the Yellow River, China. Human activity returned the river to the original channel, but renewed flooding in 1937 led to a permanent avulsion in the same area (Todd and Eliassen, 1940).

Other causes of avulsion are relegated to Group 4 (Table 3.3). Animal trails (Table 3.3:4a) can lead to overbank spill and avulsion (McCarthy *et al.*, 1992). On the Okavango fan, flood flow follows hippopotamus trails, resulting in scour, and the eventual relocation of the channel (in combination with vegetative blocking described above). In this particular setting, "hippo highways" appear to have a significant control on the location of avulsion courses.

Lateral shift of a channel until it intersects an adjacent, steeper abandoned or functioning channel may lead to capture (Table 3.3:4b) and avulsion of the main channel. In a similar manner, secondary channels formed on alluvial fans, deltas, and alluvial valleys by surface runoff and/or groundwater sapping may eventually capture the main channel and divert it into the steeper captor stream course (Denny, 1965, p. 58). This process, which differs from the others genetically because it requires the interaction of two separate drainage systems, may be considered a separate type of avulsion.

Smith *et al.* (1989) and Knighton and Nanson (1993) indicate that an avulsion can lead to anastomosis as water spreads over the adjacent floodplain. However, after a period of multiple channels, one may become dominant, and the river reverts to a single channel.

In summary, there are a variety of causes of avulsion. As a consequence, avulsions are common in many different settings (see examples in Richards *et al.*, 1993). Most avulsions can be grouped according to causal process or event that leads to instability: (1) increased sinuosity causing a decrease in gradient of the channel, (2) increased avulsion course slope due to increase in gradient away from existing channel, and (3) non-slope related reduction in the capacity of a channel to carry all the water and sediment delivered to it. For example, increased sinuosity or delta extension may combine with increased alluvial ridge relief to simultaneously decrease existing channel slope and increase lateral slope. Avulsions from channel blockage alone are probably rare, and may require that other processes provide a slope advantage for the avulsion course. An avulsion obviously will change the character of a river. As in the Australian example where a meandering channel becomes straight for a time (Figure 3.8).

Metamorphosis

Avulsion as well as other factors can lead to river metamorphosis (Table 3.1), which is not simply a pattern change. Rather it is a complete change of river morphology (Schumm, 1977, p. 159). As the word indicates, this consists of significant changes not only in the dimensions of the river, but in its pattern and shape. Considering the types of channels identified, it is possible to consider six types of river metamorphosis (Table 3.1): a straight channel changes to meandering or to braided, a braided channel changes to meandering or to straight, and a meandering channel changes to straight or to braided. It is not necessary to define each type of metamorphosis, as the change is obvious based on pattern alone.

Non-regime channels adjust to a variety of causes (Table 3.1) and the impacts of the changes can be dramatic and costly. The examples of incision and avulsion illustrate how, when rivers are near a threshold of instability, they can alter their morphologic characteristics, as a result of what can be a relatively moderate flood or human intervention.

Part II

Upstream controls

Upstream controls affect the type of river that exists downstream (Figure 1.2). They are the quantity and type of discharge and sediment load that is delivered downstream which, of course, reflects history and determines the morphology and dynamics of the downstream river.

Chapter 4

History

It may be difficult to view history as an upstream control (Figure 1.2), but changing relief, and climate through time, should have the greatest effect on upstream high relief areas, thereby increasing the impact downstream, as the delivery of water and sediment from upstream causes a downstream channel response.

The Davis cycle of landscape change through time, which is based upon the assumption of rapid uplift and then tectonic stability, can be used to explain the variability of rivers in the context of geomorphic history. Assume an uplifted block with a well-defined drainage divide and base level, and flowing water that will initiate channel incision (Figure 4.1). One can envision a large river incising down the right side of the block with the tributary incising to keep pace. The longitudinal profile of the tributary will change (Figure 4.1), with regime developing only at about stage 7 and extending upstream at stages 8 and 9 (Schumm, 1956; Whipple et al., 2000).

The hypothetical cross-sections in Figure 4.2 reveal how the valley develops through time, and they also show how, with each increment of incision, the valley walls provide a significant yield of sediment to the channel, which probably is subjected to alternate periods of aggradation and degradation.

During stages 1–4 the channel is confined by bedrock valley walls. During stages 5–7 the channel is constrained by bedrock valley walls and terraces of older alluvium. During stages 8 and 9 the channels are in regime. Hence, the distribution of the channel reaches identified by Montgomery and Buffington (1997) will depend upon the relative age of the reach, and at stages 7, 8, and 9 (Figure 4.1), the upstream reaches largely determine the character of the regime reaches downstream.

Indeed, climatic, tectonic, and base-level changes of the Quaternary Period still exert a significant impact on modern rivers, and in some cases they can determine river type. For example, Dury (1964) recognizes that underfit streams, those streams that are "too small for the valleys in which they flow," reflect an adjustment from past conditions. That is, discharge has been reduced and a large channel has become small. Dury (1964) recognizes six combinations of valley and

Figure 4.1 Longitudinal profiles of a stream incising into an uplift block (see text for explanation). Dashed line is location of cross-sections (Figure 4.2).

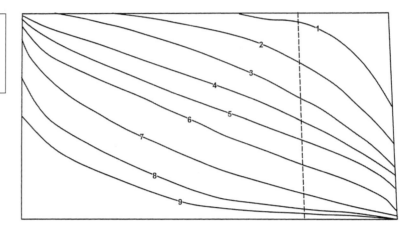

Figure 4.2 Cross-sections of valley and incising stream at times 1–9 as shown on Figure 4.1.

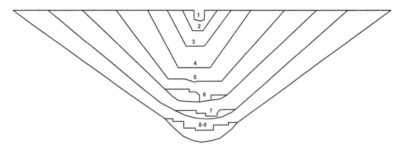

underfit streams (Figure 4.3). Two conditions exist in the examples of Figure 4.3. The first condition involves channels that maintain the pattern of the valley (example 2), and those that maintain the pattern of the paleochannel (examples 4, 6a, 6b). Therefore, although channel width and depth have decreased, the gradient remains the same. In the second condition (examples 1, 3, 5a, 5b) the sinuosity of the new channel has increased, which in turn reduces gradient. In the first condition it is probable that, although discharge was reduced, sediment loads remained proportional to discharge and, therefore, gradient was unchanged. However, in the second condition both water discharge and sediment load are significantly reduced. Therefore, in the second condition, valley slope was too steep for the new hydrologic and sedimentologic conditions, and the river meandered in order to reduce its gradient.

Evidence of large braided rivers changing to smaller meandering rivers during the climate change of the Pleistocene is widespread from eastern Europe to the Gulf Coast of the USA to Australia. Modern rivers draining from glaciers (Church and Gilbert, 1975) and paraglacial areas (Church and Ryder, 1972) show how many late Pleistocene rivers appeared before climate change produced a different type of river. Descriptions of channel change as sediment loads change with the addition of suspended-sediment load (Schumm and Khan, 1972;

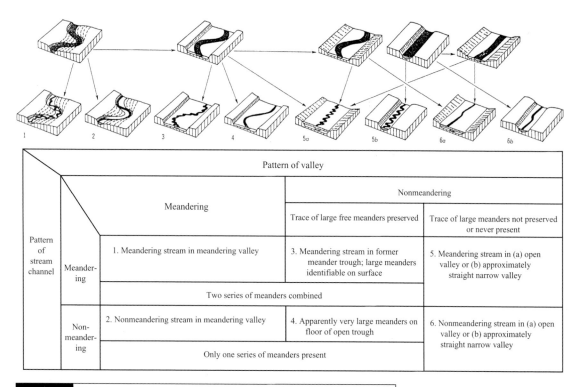

		Pattern of valley		
		Meandering	Nonmeandering	
			Trace of large free meanders preserved	Trace of large meanders not preserved or never present
Pattern of stream channel	Meander-ing	1. Meandering stream in meandering valley	3. Meandering stream in former meander trough; large meanders identifiable on surface	5. Meandering stream in (a) open valley or (b) approximately straight narrow valley
		Two series of meanders combined		
	Non-meander-ing	2. Nonmeandering stream in meandering valley	4. Apparently very large meanders on floor of open trough	6. Nonmeandering stream in (a) open valley or (b) approximately straight narrow valley
		Only one series of meanders present		

Figure 4.3 Block diagrams of underfit streams, showing character and possible origin of some combinations of stream channel and valley patterns (from Dury, 1964).

Richards, 1979; Nadler and Schumm, 1981) confirm the conclusions presented above. For example, on the Gulf Coastal Plain of the USA, there are large paleomeanders of the late Pleistocene age on the Deweyville terrace (Gagliano and Thom, 1967; Saucier and Fleetwood, 1970; Alford and Holmes, 1985), which are in striking contrast to the much smaller modern river patterns.

Elsewhere in the world, similar patterns of channel change emerge with braided streams changing to large meandering streams and then to smaller modern channels. The Riverine Plain of New South Wales, Australia, provides an example from an unglaciated region where tectonics and sea-level changes were negligible. Climate change from relatively dry to wet to intermediate conditions is reflected in the changing patterns of the paleochannels (Figure 4.4). The oldest paleochannel, which formed under a relatively dry climate, was braided and it transported large amounts of sand. It was a bed-load channel in contrast to the mixed-load younger paleochannel and the modern Murrumbidgee River, which formed under a more humid climate (Schumm, 1968).

The Polish Plain also provides an example of this type of change (Froehlich et al., 1977; Mycielska-Dowgiallo, 1977; Starkel, 1983) in even more detail, as the Pleistocene braided rivers changed to

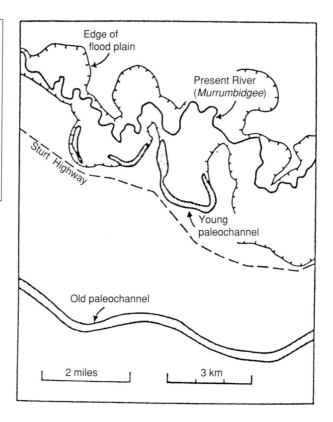

Figure 4.4 Riverine plains near Darlington Point, New South Wales, Australia. The sinuous Murrumbidgee River flows to the west (left) at the top of the figure (upper arrow). The irregular floodplain contains large meander scars and oxbow lakes (young paleochannel). The oldest paleochannel (lower arrow) crosses the lower part of the figure (from Schumm, 1968).

meandering (Figure 4.5). A transition from large meandering channels to the present river conditions can be documented by a decrease in size of the meanders that were preserved by cutoffs, although in some areas the modern rivers show a tendency to braid again as a result of agricultural activities, deforestation, and other factors.

An understanding of river variability and response to altered conditions makes it possible to speculate about river change during the period of continental deglaciation. Because of the high sediment loads and perhaps high flood peaks (high stream power), the oldest streams were wide, shallow, steep, braided bed-load channels (Figure 4.6a). As sediment loads decreased, perhaps more rapidly than discharge, a meander-braided transition pattern developed with a well-defined single thalweg (Figure 4.6b). The thalweg, in turn, became the channel as a new floodplain formed, and the channel narrowed with further reduction of sediment load (Figure 4.6c). Finally, as bed load became a fraction of its former volume, a meandering mixed-load channel with large meanders formed (Figure 4.6d). With a further significant reduction of bed load and discharge, a highly sinuous channel with low width–depth ratio formed with a multiphase meandering pattern, as smaller wavelength meanders were superimposed on pattern (d) (Figure 4.6e).

The results of studies of five Polish rivers show that the change from braided to meandering took place between 13 000 and 9 000 years BP (Figure 4.7). This example indicates how variable river

Figure 4.5 Paleochannels in the Prosna valley at Mirkow, Poland: 1, high Pleistocene terrace with braided-channel pattern; 2, and 3, slope to valley floor; 4, modern Prosna River; 5, braided pattern on valley floor; 6, paleomeanders; 7, point bars (from Kozarski and Rotnicki, 1977).

response to climate change can be. Alluvial fans can also show the same delayed response to changes of climate and sediment yield with fans in the same area of Idaho, USA, being deeply incised, shallowly incised, or not incised, although climatic and tectonic influences were the same (Schumm *et al.*, 1987, pp. 343–350).

Knox (1983) reviewed the response of rivers to the changing Pleistocene–Holocene climate for four vegetation regions ranging from the humid eastern USA woodlands to the southwestern desert shrublands, and he found a great deal of variability of channel response (Figure 4.8), as was reported for European rivers by Vandenberghe (1993).

The major difference between the humid eastern woodlands and Midwest and the dry Great Plains and Southwest appears to be in the lateral shift of the eastern streams and the cut-and-fill episodes in the western valleys. Apparently the abundant runoff of the humid eastern area could transport much of the sediment out of the valleys, whereas in the West, sediment storage and then flushing were characteristic, as alluvium was deposited and then remobilized at a later time. The importance of the above examples is that rivers in the same climatic and geologic situation can be out of phase.

Figure 4.6 Sequence of channel changes as water discharge and sediment loads decrease: (a), braided channel; (b), transitional meandering-braided channel with well-defined thalweg; (c), low-sinuosity channel; (d), relatively narrow and deep moderately sinuous channel; and (e), multi-phase meandering channel.

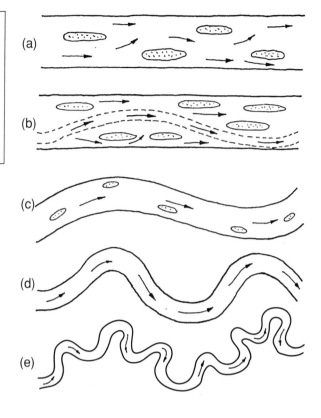

Figure 4.7 Change of river-channel patterns on the Polish Plain at the Pleistocene–Holocene transition. The straight line represents braiding, curved lines represent meandering, the dotted line represents the probable river condition. The cross-hatched zone represents the period of river change on the Polish Plain (from Kozarski and Rotnicki, 1977).

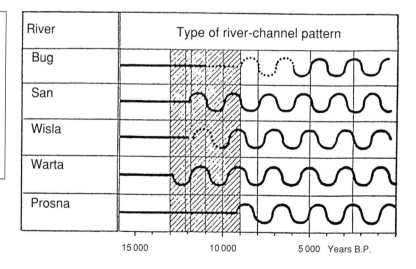

Of great importance is the establishment of valley slope as a result of past conditions. A fine example is the Mississippi River, which is one of the most intensively studied rivers of the world (Schumm *et al.*, 1994). The pioneering work of Fisk (1944) and his colleagues provides an unexcelled example of the effect of climate and sea-level changes (history) on river morphology. The two factors that have determined

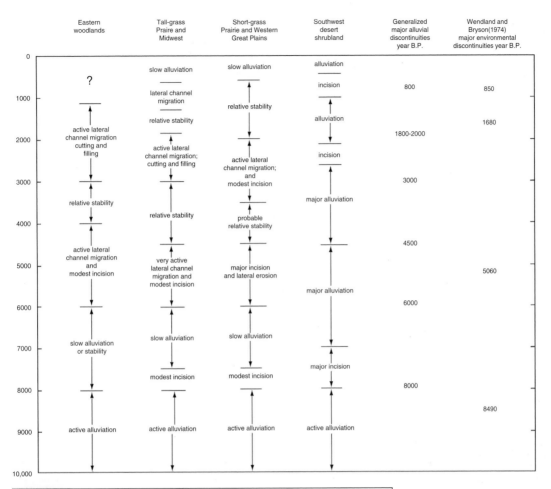

Figure 4.8 Regional alluvial chronologies. This illustration is a very generalized representation of Holocene fluvial activity derived from many site-specific alluvial sequences. Alluviation dominated during much of the early Holocene, but degradation, apparently associated with occasional episodes of intense lateral channel migration, probably dominated during the late Holocene, except in the Southwest Desert Shrubland (from Knox, 1983).

the behavior of the Mississippi River since the maximum extent of the last glaciation are the presence of an ice sheet, which supplied tremendous quantities of meltwater and sediment to the Mississippi valley, and the lowering of sea level to a maximum depth of about 130 m about 15 000 years ago. The fall of sea level lowered the base-level of the Mississippi River, but Saucier (1981, 1994) has demonstrated that the effect of base-level lowering influenced only the lower 322 km of the river.

When the continental ice sheet began to waste away, the introduction of water and sediment into the Mississippi valley was greatly increased at the same time that sea level rose between 14 000 and about 4000 years ago. Deposition in the valley occurred, and the

Figure 4.9 History of Mississippi River changes at close of the Pleistocene (from Fisk, 1944).

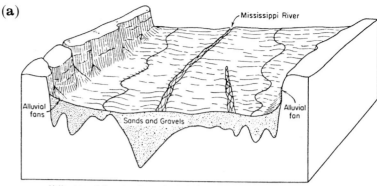

(a)

Valley aggradation stage I – sea level 100 feet lower than present
Valley slope ∼ 0.75 feet per mile

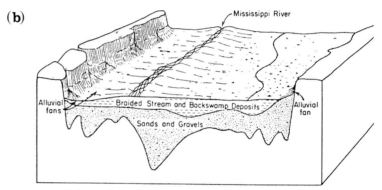

(b)

Valley aggradation stage 2 – sea level 20 feet lower than present
Valley slope 0.68 feet per mile

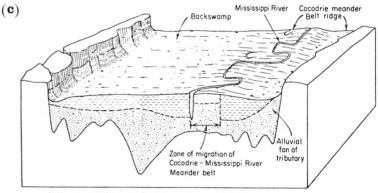

(c)

Valley aggradation stage 3 – sea level at present elevation.
Valley slope ∼ 0.60 feet per mile.

sediments presently grade upward from coarse sands and gravels through sand and silt. The lack of fine sediment (silts and clays) in the lower alluvium indicates that they were transported to the sea. According to the evidence of these deposits, the Mississippi River (Figure 1.3) at this time was a braided stream that shifted across the alluvial valley on a slope of about 0.75 feet per mile (c. 14 cm/km) (Figure 4.9a).

With continued but slower rise of sea level and a decrease in the size of the sediments moving from the north and from the tributary valleys, the sediment load was eventually reduced to fine sands, silts, and clays. According to Fisk (1944, p. 22), "The braided streams, during this stage, wandered widely on the plain and built low alluvial ridges of sands and silts . . ." The basins created between alluvial ridges received silts and clays contributed by floodwaters (Figure 4.9b). As valley alluviation continued, the gradient of the valley was reduced to about 0.60 feet per mile (c. 11.2 cm/km) and the size and amount of sediment contributed to the river decreased. At this stage, the river was flowing on a slope only 0.15 feet per mile (c. 3.1 cm/km) less than when it was braided. With essentially constant sea level and reduced sediment load, the river began to meander (Figure 4.9c). According to Fisk (1944, p. 23) the change from a braided stream to a meandering one brought about the confinement of the Mississippi flows into a single deep channel. No longer did the river wander "freely in shallow channels across the alluvial surface as did the braided stream."

This outline of Mississippi River history is similar to that suggested for the rivers of the Great Plains to the extent that they too transported larger quantities of coarser sediment and apparently reduced their gradient by developing a sinuous course. The history of the Mississippi River strongly supports the idea that meandering is largely the result of a river's attempt to reduce its gradient in response to a changed hydrologic regimen.

The River Nile, although undoubtedly affected by climate change, reflects the impact of a major ancient base-level change. It flows 6825 km from central Africa to the delta making it the longest of all rivers. It is, of course, the reason for the existence of Egypt and the great Egyptian civilization that flourished along its banks. It is an exotic river in Egypt with no perennial-flow tributaries, but ephemeral-flow wadis contribute sediment and water infrequently. According to the rules of hydraulic geometry, the channel should become smaller downstream as discharge is lost to overbank storage, irrigation diversions, and evaporation. In addition, the river should maintain a uniform pattern, and gradient should decrease slowly or remain constant downstream. However, this is not the case as width, gradient, and channel pattern vary significantly (Schumm and Galay, 1994). Nevertheless, the river is straight in contrast to other large rivers (Figure 4.10), and because ancient Egyptian temples occupy its banks, it appears to be relatively stable. The Nile today has been described as a "very low energy river with little capability to erode its banks (only 12% of its banks are experiencing erosion) or change its channel" (Mercer, 1992). Hence, it is very unlike the Mississippi River, which, of course, has a very different history.

The River Nile has a fascinating history, which includes the integration of White and Blue Nile drainage with the lower Nile and deep incision associated with desiccation of the Mediterranean Sea (Hsü, 1983; Said, 1981). The Nile formed a deep canyon about 10 million years ago as the Mediterranean Sea evaporated (Hsü, 1983).

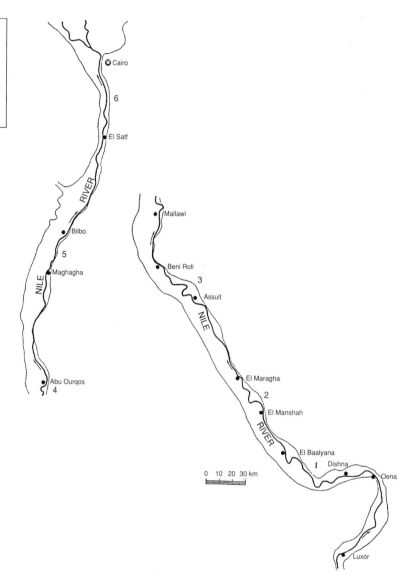

Figure 4.10 The Nile downstream from Luxor. Numbers identify geomorphic reaches (from Schumm and Galay, 1994). Sinuous reaches 2 and 3 are the result of faulting, which locally steepened the valley floor.

North of Cairo, there is a buried canyon, the bottom of which is 2500 m below sea level. Rise of sea level 5 million years ago caused marine waters to enter the canyon and to reach as far south as Aswan. The Nile deposited fluvial sediments at the head of this gulf and progressively filled the valley to the north. For more detail on the history of the Nile during the past 40 000 years, see Butzer (1976, 1980).

The filling of the Nile-canyon estuary was much like delta building. The river constructed a valley slope that was needed only for the transport of its water and sediment load. During humid periods, wadis contributed sediment that steepened the valley floor locally, but there was no great influx of glacial outwash sediments to steepen the entire valley. As a result, the Nile is a remarkably straight river

(Figure 4.10) that flows on a relatively gentle valley slope. The valley slope is actually equal or steeper than that of the Mississippi valley, but the Nile discharge prior to the construction of the High Aswan Dam was half that of the Mississippi. Obviously, during the last 5000 years the river has aggraded and degraded as climate and hydrologic conditions changed, but the relatively gentle valley slope was maintained (Schumm and Galay, 1994). These conditions were very different from those of the Mississippi River where the influx of glacial outwash from the north and lowered sea level to the south produced a valley slope steeper than that required by the modern Mississippi River, which to compensate, developed a meandering course.

Church (1981) recognized that the response of a river reach to climate or base-level change "depends greatly on position in the channel system and on the recent local landscape history." For example, if the shape, gradient, and dimensions of a channel adjust to discharge and sediment load, why is the valley gradient too steep for the discharge and load of sinuous streams? Sinuosity reflects not only the ratio of channel length to valley length but also the ratio of valley slope to channel slope. In some cases, valley gradient is 2.5 times as great as stream gradient, whereas in other instances valley and stream gradients are almost identical. If tectonic factors have not modified the slope of a valley, the gradient of the alluvial surface should be at just that slope required for the movement of the water-sediment mixture through the valley. The fact that the surface of the alluvium is too steep in many cases requires an explanation. That explanation depends on an understanding of the history of changes in stream regimen following the Pleistocene. Most rivers flow on the upper surface of alluvium which fills valleys cut into bedrock. The valleys and the associated alluvium are a result of uplift and the changes in base-level, climate, and runoff during and following the Pleistocene. Hence, history can be of vital importance.

To complicate matters, the existence of glacial lakes in valleys result in a very flat slope over which a river must flow. Usually the flat slope requires a straight channel. The Illinois River provides such an example, and Lewin (1987) describes the changes in the lower Vyrnwy River, a tributary to the river Severn in Great Britain, as it enters a reach of flat gradient on an ancient lake bed. The channel is stable, the width–depth ratio decreases and as expected, sinuosity decreases significantly from 1.80 to 1.25 on the flat valley floor.

History can be very important in determining river type and reach variability, depending largely on the history of climate and hydrologic change. Other factors that determine valley gradient, in turn, can determine modern river morphology and variability of the river from reach to reach.

Chapter 5

Tectonics and relief

Tectonics is an important determinant of river type (Figure 1.2). It generates the relief that drives the erosional machine; it generates earthquakes, which provide vast amounts of sediment to a river; and it causes avulsion and gradient changes. All of the above influence river type and behavior. Significant effects of tectonics are visible in the Amazon basin and on the Hungarian Plain (Schumm et al., 2000), and as an upstream control its downstream impacts can be significant. The effects of local, active tectonics, which impact limited reaches of a river, are discussed in Chapter 11.

High relief mountains, that are tectonically active and that were or are glaciated produce large quantities of coarse sediment, which in turn, produce wide, high width–depth ratio, braided bedload rivers. In contrast, low relief, stable mountains, especially in humid regions, produce low width–depth ratio, sinuous or straight channels.

As is generally the case, the effect of relief and slope on sediment yields and hence upon the type of river found downstream is modified by other variables, such as lithology and climate (Figure 1.2). Ahnert (1970) collected data for 20 European and US river basins with a range of relief from 89 m to 2869 m. He concluded that there is a direct correlation between relief and rate of erosion and sediment yield. In contrast, for small drainage basins in western, semiarid US the relation is exponential (Figure 5.1). Perhaps Ahnert's data reflect more precipitation with more vegetation, less land use, and more resistant rock at higher elevations, which produce less sediment than might be expected. For Japan, Ohmori (1982) reports that both power and exponential relations between sediment yield per unit area and basin slope are highly significant. In any case, steep, high relief drainage basins will produce more sediment and steep active channels.

This type of variability can be expected because the data are obtained from drainage basins of different lithology and climate, and the sizes of the basins are not constant (Summerfield and Hulton, 1994; Milliman and Meade, 1983; Ludwig and Probst, 1996; Dedkov and Moszherin, 1996; Jansson, 1988). Nevertheless, if all other variables are held constant, increased relief and slope will produce more sediment and influence river type (Figure 5.2).

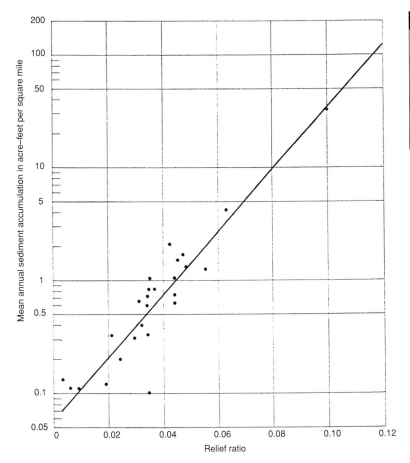

Figure 5.1 Relation between mean annual sediment accumulation in small reservoirs and relief ratio for drainage basins on Fort Union formation in the USA. Relief ratio is the ratio of maximum relief to length of a drainage basin (from Hadley and Schumm, 1961).

Schumm (1963) has argued that uplift on average is almost eight times the rate of average denudation. Of course, uplift must exceed denudation or mountains would not exist. Nevertheless, rates of denudation in steep terrain can equal uplift because mass movement becomes a dominant process, and a balance between uplift and denudation does pertain in the New Zealand Southern Alps and in Taiwan. Active tectonic areas will, of course, provide large amounts of sediment, much of which can be the result of mass movements (Brunsden et al., 1981; Hovius et al., 1998; Schmidt and Montgomery, 1985; Hovius, 1998). Earthquakes modify landforms by landsliding, and they can significantly affect river morphology and hydrology temporarily or permanently by triggering an avulsion.

Many braided New Zealand rivers show the effects of huge catastrophic landslides, as indeed do rivers in other mountainous areas of the world (Hewitt, 1998). High relief is a pre-requisite, but a triggering mechanism also is required. Triggering mechanisms may include heavy rainfall or large earthquakes, both of which can produce clusters of landslides. Hicks et al. (1996) suggest that the highest rates of fluvial sediment transport in New Zealand tend to cluster along the

Figure 5.2 Variation of sediment load with drainage basin area for seven topographic categories of river basins. Note the marked difference between high mountains (a, b, c, d), upland (e), and lowland and coastal plain basins (f, g) (from Milliman and Meade, 1983).

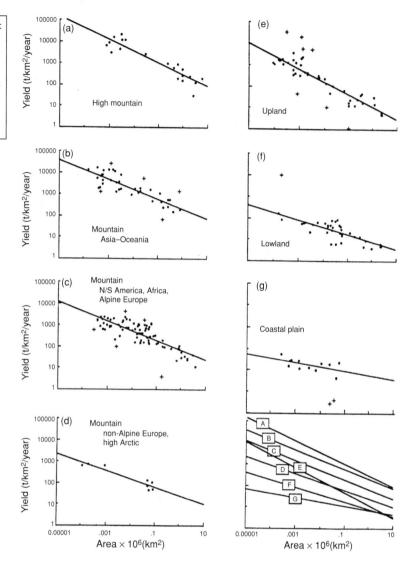

boundary between the Pacific and Indian tectonic plates, that is, they are related ultimately to rates of tectonic uplift and relief creation.

Rivers often follow structural lows or major geofracture systems. Melton (1959) estimated that between 25 percent and 75 percent of all continental drainage in unglaciated regions has been tectonically influenced or controlled, while Potter (1978) concluded that some large rivers have persisted in essentially their present locations for hundreds of millions of years, because they occupy major tectonic zones. Tectonic deformation by altering channel gradient can have a significant impact on the behavior and form of a particular reach of river.

Earthquakes that occur a considerable distance upstream may affect the downstream river channels. For example, the 1950 earthquake in Bangladesh caused massive landslides in the Himalayas

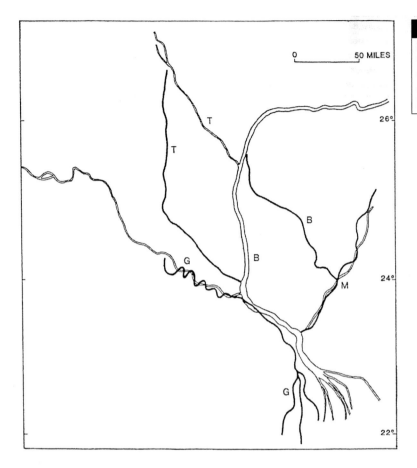

Figure 5.3 Sketch map showing positions of Brahmaputra River (B), Megna River (M), Tista River (T), and Ganges River (G) in 1779 (solid lines) and at present (double lines).

which caused significant aggradation in the upper reaches of the Brahmaputra River. The long-term effects could be considerable, as this sediment moves downstream. For example, the 1897 earthquake, that was centered on the Shillong Plateau, caused widespread damage across Bangladesh (Molnar, 1987). The Brahmaputra River in Bangladesh avulsed 80 to 100 km to the west in the late eighteenth century during a time of large earthquakes (Figure 5.3). The earthquakes may have played a role in this avulsion, but tilting (Morgan and McIntire, 1959; Johnson and Alam, 1991) and subsidence or simply the selection of a more direct route to the sea could have caused the avulsion.

There appear to be major changes of dimensions of the Brahmaputra River during historic time. For example, Rennell (1765) found that it was 4 to 5 miles (c. 6–8 km) wide above the Ganges junction. Bristow (1987) however, states that the present mean width is 10 km. Also, the Ganges on the Rennell map has a very sinuous reach, whereas now it is much straighter (Figure 5.3). Coleman (1969) notes that the oldest courses of the Brahmaputra River were sinuous. Perhaps increased sediment loads, that result from earthquakes, have altered the dimensions and pattern of the rivers during the past 200 years.

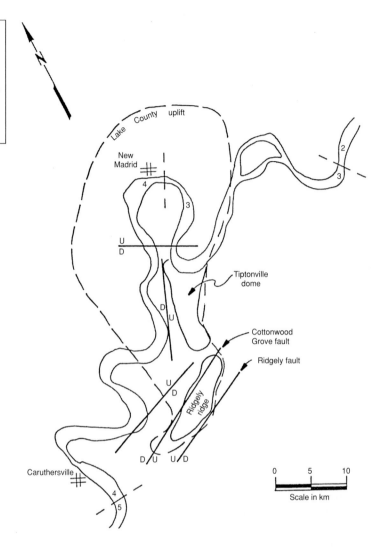

The history of the rivers of Bangladesh is complicated by discharge and sediment loads derived from the Himalayas, sea level change, subsidence of the delta, and local tectonics. Geologists conclude that the alluvial plains of the Brahmaputra–Ganges delta are being actively deformed under complex stress caused by collision of the northeast corner of the Indian tectonic plate with Asia. Warping and faulting elevate some areas and depress others. Elevation differences, although small, are critical in controlling the pattern of inundation during seasonal flooding. The distribution, flow direction, depth, and persistence of floodwater on the alluvial plain are also indications of Holocene subsidence and uplift. The history of avulsions on the delta also support the conclusion that the delta is being affected by active tectonics.

Between December 16, 1811, and March 15, 1812, a total of 203 damaging earthquakes occurred in the New Madrid (Missouri) seismic zone (Figure 5.4). Between December 1811 and February 1812, three earthquakes exceeded magnitude 8. The effects of the New Madrid earthquakes on the Mississippi River have been described by several eyewitnesses because boat traffic on the river was heavy during the winter of 1811–12. Summaries of these accounts are contained in published documents by Fuller (1912) and Penick (1981). For example, in a letter dated February 18, 1812, J. Smith described how the river had "wholly changed" between New Madrid and Memphis (Tennessee). In this reach, the development of bars and lodging of snags created severe hazards to navigation (Penick, 1981). There is a recurring description of two distinct "rapids" in the Mississippi River near New Madrid. Penick (1981) concludes that they formed during the February 7, 1812 earthquake with its epicenter near New Madrid.

Another consistently described effect of earthquakes on the Mississippi River is extensive bank failure. Fuller (1912, p. 89) recorded the observations of Bradbury the English naturalist, who stated, "immediately the perpendicular surfaces (banks) both above and below us, began to fall into the river in such vast masses as nearly to sink our boat by the swell they occasioned." Bank erosion was also described by Lloyd (Fuller, 1912, p. 89): "At times the waters of the Mississippi were seen to rise up like a wall in the middle of the stream and suddenly rolling back would beat against either bank with terrific force . . . during the various shocks the banks of the Mississippi caved in by whole acres at a time."

Numerous islands within the channel were reported to have sunk during the earthquakes. Fuller (1912, p. 10) writes of the December 16, 1811, earthquake in the following manner: "On the Mississippi, great waves were created, which overwhelmed many boats and washed others high upon the shore, the return current breaking off thousands of trees and carrying them out into the river. High banks caved and were precipitated into the river, sand bars and points of islands gave way, and whole islands disappeared."

Eye witness accounts of the effects of the earthquakes are consistent in describing tremendous numbers of bank failures that introduced large quantities of sediment into the river. Walters (1975) and Walters and Simons (1984) assumed that this great increase in sediment load could affect the river far downstream, and they note that bank caving was described from the Ohio River almost to Memphis. This undoubtedly caused widening of the river, and indeed, for 190 miles (c. 304 km) downstream of Cairo (Illinois), channel width was well above average, whereas downstream it was about or below average (Figure 5.5).

In addition, the number of cutoffs increased after the earthquake (Figure 5.6) from four in the 52-year period, 1765 to 1817, to eight in the 56-year period between 1818 to 1874, after the earthquakes. The Mississippi River example provides a clear and convincing case of drastic and long-term effects of earthquakes on alluvial rivers.

Large alluvial rivers that flow down wide alluvial plains can avulse as a result of faulting and earthquakes. For example, the Indus River in Pakistan is flowing on an alluvial ridge of its own making

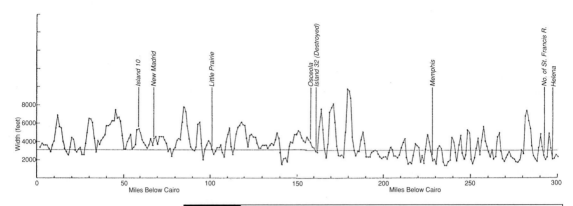

Figure 5.5 Width of Mississippi River in 1821 between Cairo, Illinois and Helena, Arkansas (from Walters, 1975). Horizontal line indicates average width (3100 feet; *c.* 945 m).

Figure 5.6 Number of meander cutoffs for four periods (from Walters, 1975).

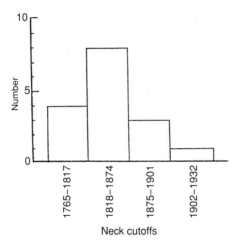

(Jorgensen *et al.*, 1993). The river has avulsed widely in the past (Figure 5.7), and it is probable that earthquakes have played a role because there are numerous active faults in the Indus Valley (Kazmi, 1979). The most spectacular effect of active faulting is in the Rann of Cutch fault zone. In this region of the lower Indus Valley in 1819, a severe earthquake resulted in the uplift of a 16-km wide and 80-km long tract of alluvial land with a relief of about 6 m. This feature was locally known as Alah Bund (Oldham, 1926). The eastern branch of the Indus was blocked by the formation of the Alah Bund (Oldham, 1926) and the channel at that time was dry, but flow was re-established during a flood in 1828.

The Indus River in Sindh, Pakistan, exhibits a range of modern channel patterns, and the wide alluvial plain displays numerous ancient courses of the river (Figure 5.7). The river changes from braided to anastomosing to meandering as it crosses several apparently active structures (Figure 5.8). Obviously, the river has

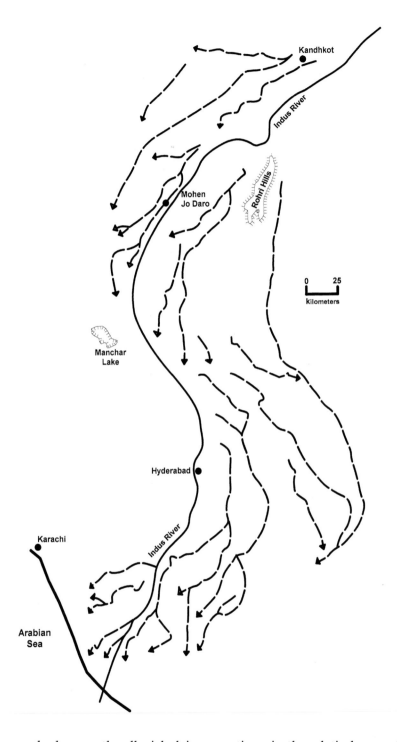

Figure 5.7 Present (solid line) and ancestral (dashed lines) courses of the Indus River, Pakistan (modified after Flam, 1993).

avulsed across the alluvial plain many times in the relatively recent geologic past. Of particular interest is the fact that there appear to be two locations where avulsion has occurred repeatedly, one is upstream of Kandhkot and the other downstream to the southeast of Manchar Lake (Figure 5.7). Both of these locations are near major

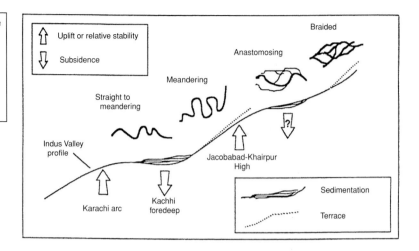

Figure 5.8 Cartoon showing the relationship of the Indus valley profile and channel pattern to tectonic elements. Note that the slope and pattern changes are exaggerated for illustration (from Jorgensen *et al.*, 1993).

tectonic boundaries (Jorgensen *et al.*, 1993; Flam, 1993), and it is reasonable to conclude that avulsion in this case is not the result of cross-valley slope variations, but that it occurs at specific sites which have tectonic controls. As a result of these controls, the modern Indus plain can be thought of as two alluvial fans, one focused at Kandhkot and the other east of Manchar Lake. Avulsion occurs at the apex of these fans. Indus River behavior suggests that the location of avulsions may be controlled by tectonics. This conclusion is reasonable because avulsions occur repeatedly at the same location.

Study of paleochannel positions on the North China Plain reveal that points of repeated diversions and locations of levee breaches are influenced by basement tectonics (Xu *et al.*, 1996, p. 34). Strong earthquakes have been recorded. There have been six earthquakes of 8.5 magnitude which could affect the Yellow River (China) among others in a similar manner to the New Madrid earthquakes that affected the Mississippi River.

In summary, the effect of tectonics on alluvial rivers can be widespread, and even the largest rivers are affected. Earthquakes and avulsion can greatly modify river morphology downstream, when bank failure and cutoffs widen and straighten channels, and, of course reach morphology will vary significantly.

Chapter 6

Lithology

Lithology obviously affects runoff and sediment yield and therefore, it is an upstream control (Figure 1.2). For a given relief, sound rock will yield less sediment, but shattered rock will cause mass movement and the delivery of large quantities of sediment to the main streams. As considered previously, the type and amount of sediment delivered to a valley can determine the type of channel found there, as well as the variable morphology and behavior of the channel.

Schumm (1960, 1961) argued that, at least for sand-bed rivers, those draining from shale or siltstone areas will have a high percentage of silt and clay in banks and bed, and they will be narrow, deep and sinuous, whereas those draining from sandy and gravel areas will be wider and shallower, and relatively straight. A good example of this is provided by the Kansas River system in Colorado and Kansas.

The Kansas River (Figure 1.3) is formed where the Smoky Hill and Republican Rivers join in central Kansas. The Smoky Hill River in its headwaters drains sandy sediment, as does the Republican River. However, the Smoky Hill River is joined by two large tributaries, the Saline and Solomon Rivers that drain from Cretaceous shales. Moving downstream from a small upstream tributary of the Smoky Hill River (Figure 6.1, Site 32), the channel is very sandy (Sites 32–37) and width increases dramatically, but depth only doubles. A significant increase of silt-clay in the channel (M) occurs between Sites 37 and 41, where the Saline and Solomon Rivers deliver silts and clays from shale outcrops. As a result, width decreases and depth increases and width–depth ratio decreases from about 33 to 13. A large tributary, the Republican River, enters from the north, downstream of site 43, delivering a very large sand load derived from Tertiary-age sandstones and conglomerates. As a result, both width and depth increase (sites 44, 45), and width–depth ratio increases to about 50.

Changes through time also reflect changes in the type of sediment load. The channel of the Murrumbidgee River in New South Wales, Australia changed from braided to sinuous, as sediment loads changed from bed load to mixed load (Figure 4.4), when climate changed from relatively dry to wet, thereby decreasing total sediment loads and especially bed load.

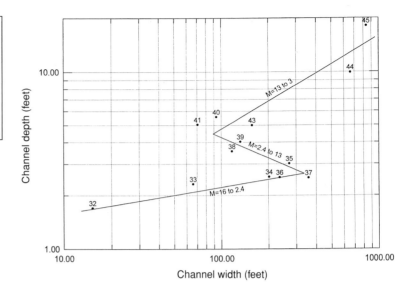

An extreme example of the effect of sediment type delivered from upstream is provided by the Citanduy River of Java (Stevens, 1994), that has 90 percent of its sediment load as silt and clay, which is derived from volcanic ash. The banks and bed of the channel are composed of clay. The result is a low width–depth channel (10) with a sinuosity of 1.7 to 2.0. The river is very stable, and it resembles Brice's uniform width channels (Figures 2.1 and 2.2). In fact, when meanders were cut off, which reduced sinuosity from 2.0 to 1.4, the river remained stable as a result of the cohesive floodplain sediments (Figure 6.2). This is very unlike the response of the mixed-load Mississippi River, which reacts significantly to the increased gradient resulting from cutoffs (Winkley, 1977).

Nevins (1965) correctly concludes that the character of New Zealand rivers "is largely determined by the materials forming its bed and banks" and therefore, head-waters rock type is important. The petrology of New Zealand leads to the identification of four geologic regions as follows: (1) hard rocks, greywacke, granite, gneiss; (2) softer schists; (3) shales and mudstones; and (4) volcanics (Figure 6.3).

According to Nevins, the hard rocks are shattered and deliver coarse gravel to the river. Therefore, they are steep, braided, and active. The schists produce weaker sediment loads and the rivers have a gentler gradient and are "mature." The shales and mudstones yield even gentler gradients and are meandering. The volcanic terrain produces a mixed sediment load of volcanic ash and gravel and therefore the channels are variable. Although Nevin's geologic divisions are very general, nevertheless he supports the contention that lithologic characteristics are very important in determining sediment loads and channel character downstream of the sediment sources.

There is little doubt that rivers draining from steep mountain regions or glaciated mountains will be braided. Of course as the

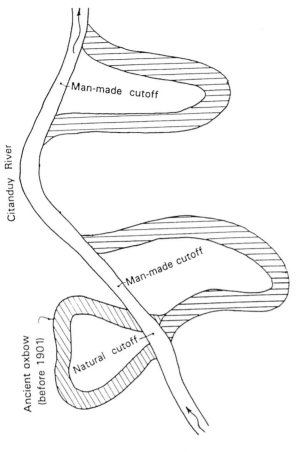

Figure 6.2 Natural and man-made cutoffs along Citanduy River, Indonesia. Dramatic shortening had little impact on river behavior (from Stevens, 1994).

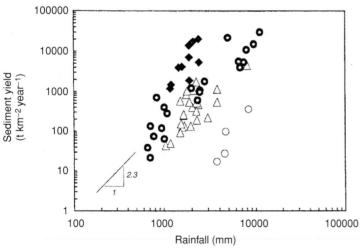

Figure 6.3 Sediment yield as a function of mean annual rainfall for New Zealand catchments with different lithologies (from Hicks et al., 1996).

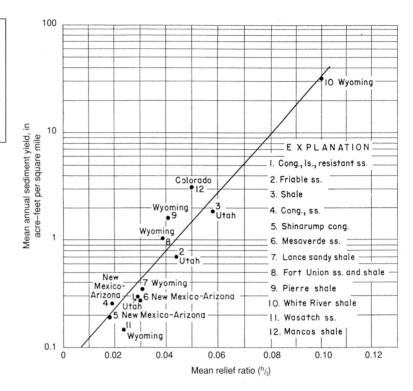

Figure 6.4 Relation of mean annual sediment yield to mean relief ratio (maximum relief (h) divided by maximum length (l) of basin) for 59 small drainage basins in Wyoming, Colorado, Utah, New Mexico, and Arizona (from Hadley and Schumm, 1961). ss, Sandstone.

sediments move downstream and are mixed, the great contrast between bed load and suspended-load streams are lost. Montgomery and Buffington's (1997) classification of river reaches (Figure 2.6) reflects not only gradient but type of sediment from boulders and coarse gravel in cascade and step pool reaches to sand and fine gravel in pool-riffle and regime reaches.

Rock resistance to erosion is an important factor determining sediment yield rates and the type of sediment delivered to and controlling stream morphology. For example, sediment yield from small drainage basins in the Cheyenne River basin of southeastern Wyoming depends not only upon the average slope of the drainage basin (relief ratio), but also on rock type (Figure 6.4). The more resistant sandy rocks (1, 4, 5, 6, 11) produce less sediment both as a result of resistance to erosion and higher permeability whereas weaker silty sandstones (2, 7) and shale and siltstones (3, 8, 9, 10, 12) produce higher fine sediment loads. The sandy, gravelly units produce bedload type channels that are wide and shallow and the shale and siltstone units produce low width–depth ratio channels.

Hicks *et al.* (1996) show how for a given amount of precipitation, sediment yield varies with rock type (Figure 6.3) in New Zealand. A similar pattern (Figure 6.5) emerges for data from Japan (Shimizu and Araya, 2001). For a given drainage area, resistant igneous rocks yield less sediment than younger sedimentary rocks. Bedload should be greater for the resistant rocks yielding a braided river.

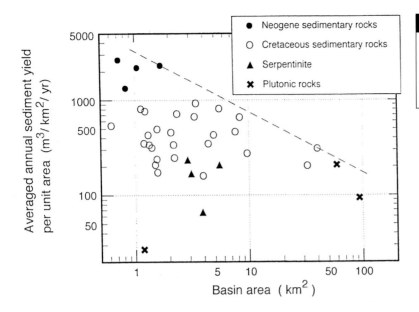

Additional examples of the effect of type of sediment load are provided by the William River in Canada (Smith and Smith, 1984) and small streams in Cornwall, England (Richards, 1979). William River is slightly sinuous and not braided. However, when it receives a major increase of aeolian sand from the Athabasca dune field, width increases fivefold and width–depth ratio increases tenfold. As a consequence, the river braids, as it would if the sand was derived from a sandy bedrock.

The Cornwall rivers drain from an area where kaolin is mined (Richards, 1979). The mining process significantly increases the clay content of the discharge, which causes a decrease of width and width–depth ratio. This result is very similar to the results of an experiment when kaolinite was added to the flow, which converted a straight channel with a sinuous thalweg to a meandering channel (Schumm and Khan, 1972). Again, the results are similar to what is expected when a river crosses a geologic boundary between sandy and clayey bedrock or between resistant and weak bedrock.

Types of bedrock exposed in the headwaters of a drainage basin greatly influence both valley and channel character downstream. The quantity and type of sediment produced determines channel morphology, and variations in type of sediment load can profoundly impact river reaches.

Chapter 7

Climate: hydrology

Climate is a controlling factor in determining river hydrology and type (Figure 1.2). Depending upon amount of precipitation, rivers will be ephemeral, intermittent, or perennial, and, of course, as the hydraulic geometry relations indicate, the more water the larger the channel. For the purposes of this discussion, it is assumed that hydrology is determined by upstream climatic conditions.

In the later nineteenth century, British engineers, in what is now India and Pakistan, were designing major canal systems. They recognized that in the artificial channels, dimensions were related to discharge. For example, Kennedy (1895) found that for his canals:

$$A = 0.8 \ Q^{0.167}$$
$$w = 2.67 \ Q^{0.5}$$
$$d = 0.64 \ Q^{0.33} \tag{7.1}$$

A is cross-sectional area of the channel and w and d are width and depth, and Q is discharge. These relations and others were used to design regime canals that did not progressively degrade or aggrade through time. They, in fact, were artificial graded rivers. Reviews of this canal design literature can be found in Leliavsky (1955) and Garde and Ranga Raju (1977). Leopold and Maddock (1953) took this approach with natural rivers and their hydraulic geometry relations showed, as expected, good correlations between discharge and channel width, and with depth, and with velocity.

All of this work led to the conclusion that in these adjustable channels, changes of sediment load and discharge would result in adjustments of channel width, depth, gradient, and pattern. For example, Lane (1955) summarized these relations by presenting a qualitative relation, including bed load (Qs) and sediment size (d_{50}) as follows:

$$Qs \cdot d_{50} \approx Q, \ S \tag{7.2}$$

A change in sediment character requires a change in Q and/or S to compensate.

Lane's equation involves only one aspect of river morphology, gradient. Obviously, if discharge or sediment load changes, other aspects

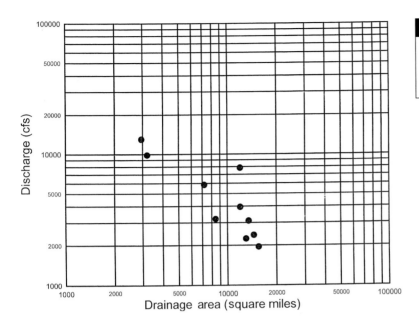

Figure 7.1 Calculated bankfull discharge of Finke River, Australia showing downstream decrease. cfs, cubic feet per second (0.0283 m³/s).

of the channel will change. For example, channel width (b), depth (d), and meander wavelength (ℓ) are directly related to discharge (Q) whereas gradient (S) is inversely related to discharge. From this, the following generalized relation is obtained:

$$Q \approx \frac{b, d, \ell}{S} \tag{7.3}$$

Either mean annual discharge (Qm) or mean annual flood (Qma) can be substituted in this and subsequent equations. Considerable independent information is available to demonstrate that the relationships expressed by Lane are valid.

Some studies (e.g. Schumm, 1968) have shown when other things are constant, that bed-material load is directly related to width, meander wavelength and slope and inversely related to depth and sinuosity.

$$Q_s \approx \frac{b, \ell, S}{d, P} \tag{7.4}$$

Obviously, if there is a change of Q or Q_s downstream, the river will be different. The downstream hydraulic geometry of Leopold and Maddock (1953) requires that with an increase of flow, channel width and depth will increase in a downstream direction, but if a river is exotic, the reverse will pertain. Downstream discharge will decrease and so should channel dimensions.

The Finke River in central Australia flows from the McDonald Range west of Alice Springs to the Simpson Desert. Rainfall is sparse in central Australia, and it decreases downstream along the Finke River with the result that bankfull discharge decreases with increasing drainage area (Figure 7.1). As a result, width (Figure 7.2) and depth (Figure 7.3) decrease downstream. Tooth (2000) demonstrates similar downstream changes for several other exotic Australian rivers. As an

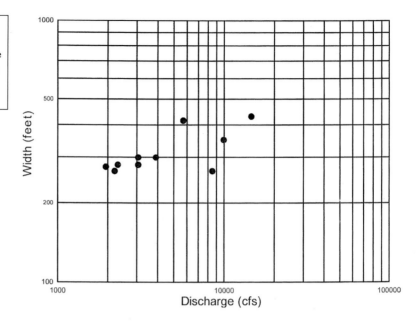

Figure 7.2 Channel width of Finke River plotted against calculated bankfull discharge. Note that downstream direction is from right to left. cfs, cubic feet per second (0.0283 m^3/s).

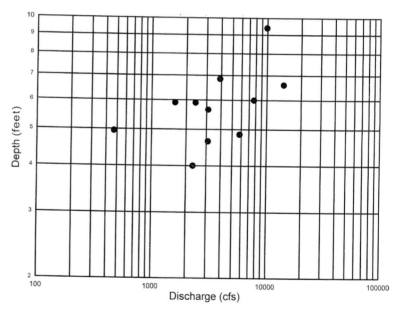

Figure 7.3 Channel depth of Finke River plotted against calculated bankfull discharge. Note that downstream direction is from right to left. cfs, cubic feet per second (0.0283 m^3/s).

indication of the infrequency of flow, residents at Idracowra Homestead in about the center of the Finke River drainage basin stated that there was one almost bankfull flood in 12 years and three lower flows in five years. As a result of the aridity, it requires 1000 square miles (c. 1600 km^2) to support 2000 head of cattle. Also characteristic of this environment is that the flows generated by precipitation in one part of the basin do not always reach downstream areas. The flow is lost largely to infiltration into the dry channel.

Climate is exceptionally important in determining river type because it establishes the hydrologic character of the drainage basins,

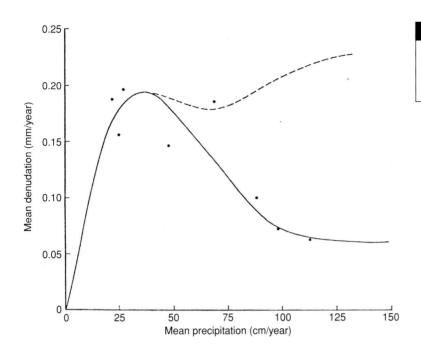

Figure 7.4 Effect of climate on denudation rate. Solid line represents conditions with natural vegetation. Dashed line represents effect of agricultural activity.

the vegetation which stabilizes hillslopes and channel banks and the quantity of sediment delivered to the active channels.

Sediment yields

Natural erosion rates vary greatly with rock type and relief within one climatic region, but they are also closely related to climatic controls. The relationship between natural rates of erosion and precipitation, for continental climates (Figure 7.4) reveal that maximum natural erosion rates, for small drainage basins between 26 and 130 km², occur in semiarid regions (Langbein and Schumm, 1958). The form of the curve is largely dependent on vegetation. Under humid conditions vegetation protects the soil, but in arid regions there is insufficient rainfall and runoff to move large quantities of sediment. Therefore, at an intermediate rate of precipitation vegetational protection is low, yet there is sufficient runoff to move sediment out of the drainage basin. The broken line extension of the curve (Figure 7.4) represents an upward shift of the right side of the curve, that results from the removal of natural vegetation by agricultural activities.

The left side of the curve is sketched with little data, although Larrone and Wilhelm's (2001) data from the Negev indicate very low sediment production in an arid landscape, and data from Indonesia supports the right side of the curve with low sediment production from the humid tropics (Cecil *et al.*, 1993). Dense tropical vegetation significantly inhibits erosion, but if the vegetation is removed, erosion will be intense. Humid and tropical regions are characterized by deep chemical weathering, whereas semiarid and arid regions are

Table 7.1 Characteristics of Rios Tonoro and Guanipa

River	Drainage area (sq/mi)	Valley slope (ft/mi)	Channel width (ft)	Sinuosity	Median sediment size (mm)	Mean annual discharge (Q) (cfs)	Maximum annual discharge (Qp) (cfs)	Qp/Q
Tonoro	500	0.8	600	1.1	0.35	400	18 900	47
Guanipa	1100	0.7	50	2.3	0.35	600	3700	6

(From Stevens *et al.*, 1975b)
cfs: cubic feet per second

characterized by physical weathering. In the former, suspended sediment loads of silt and clay will dominate and in the latter, sand and gravel bed loads will dominate, which suggests that braided rivers will be common in semiarid and arid areas and meandering and straight rivers will predominate in humid and tropical climates.

For the same type of sediment load, discharge determines the dimensions of a channel. In addition, the nature of discharge is very influential. For example, one such hydrologic variable is peak or maximum discharge. Single floods can have significant effects on channels. For example, Wohl (2000b) lists a variety of channel changes as follows: bed erosion, bank erosion, deposition within the channel, growth of bars and islands, floodplain accretion, channel lateral shift, and pattern change. However, with time, the channel can recover and revert back to its normal morphology.

It is rare to find streams that drain geologically similar areas and yet have very different flood peaks, but two such rivers are the Rios Guanipa and Tonoro in northeastern Venezuela (Stevens *et al.*, 1975b). A comparison of the morphology and hydrologic character of both rivers is presented in Table 7.1. The major differences in width and sinuosity (ratio of channel to valley length) appear to be the result of the great difference in flood characteristics.

Although the records are short, they indicate that rivers with high ratios of peak to mean discharge are morphologically different from rivers with low ratios. In a general way this is substantiated by Gupta (1975) for two rivers in Jamaica (Table 7.2). In this case the only factor that can explain the difference between the braided Yallahs River and the narrower, more sinuous Buff Bay River is the marked seasonality of precipitation in the Yallahs River drainage basin. Annual precipitation is similar in both drainage basins, but larger floods occur in the braided Yallahs channel.

Formerly meandering rivers can be converted to straight channels by a combination of high peak discharges and an influx of coarser sediment. For example, the highly sinuous, relatively narrow and deep Cimarron River (Figure 1.3) channel of southwestern Kansas was destroyed by the major flood of 1914 (Schumm and Lichty, 1963). Between 1914 and 1939 the river widened from an average of 50 feet

Table 7.2	Comparison of the Yallahs and Buff Bay rivers	
Factors	Yallahs	Buff Bay
1. Channel pattern	Braided or straight	Meandering or straight
2. Channel shape	Wide shallow flat-bottomed channel with steep banks	Deep round-bottomed channel with sloping banks
3. At-a-station change in channel with rising discharge	Rate of change in width with increasing discharge greater than simultaneous rate of change in depth	Rate of change in depth with increasing discharge greater than simultaneous rate of change in width
4. Depositional features	Mid-channel bars and sidebars	Point bars and large, high, vegetated islands in lower reaches
5. Bed material	Coarse and at least half of the river bimodal. Principal mode – pebbles and cobbles. Secondary mode – silt or sand (when present)	Coarse and bimodal. Principal mode-pebbles or cobbles. Secondary mode-sand.
6. Bank material	Similar to bed material in texture. Weak fining upward sequence at places. Locally imbrications in pebbles	Very shallow layer of silt and sand on top. Rest similar to bed materials. No significant structures observed.
7. Floodplain height above channel bed	90 to 100 cm	90 to 180 cm

(c. 15 m) to 1200 feet (c. 366 m), and the entire floodplain was destroyed. Large floods moved considerable sand and caused this transformation despite the fact that annual discharge was probably less during the drought of the 1930s. The hydrologic record is short, but there was an increase in annual discharge after 1940, and the channel narrowed to 200 feet (c. 61 m).

Precipitation data indicate that the years 1916–41 were generally a period of below-average precipitation. Thus, during years of low runoff and high flood peaks, the Cimarron River was converted from a narrow sinuous channel characterized by low sediment transport to a very wide, straight braided bedload river. Similar changes on the Gila River in Arizona have been documented by Burkham (1972) and for the Macdonald River in Australia (Erskine, 1986a).

Although the result of human modification of discharge, the Platte River (Figure 1.3) in Nebraska is a good example of how the change from intermittent to perennial discharges affects this river flowing from semiarid Wyoming and western Nebraska to subhumid eastern Nebraska (Figure 7.5). This system will be considered in some detail because of the nature of the hydrologic impact upon the river and because it provides information about the transformation of both a braided and an anastomosing river.

The Platte River system has been affected by the impact of dams and reservoirs and especially irrigation diversions and irrigation

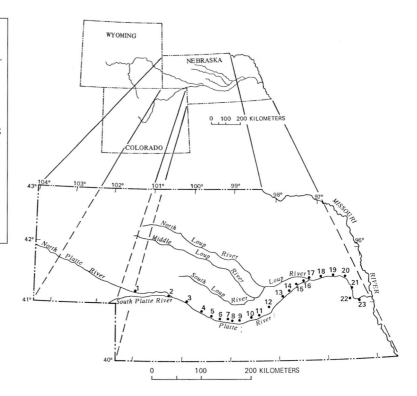

Figure 7.5 Map showing location of towns along North Platte and Platte River in Nebraska. Key to numbered towns: 1, Keystone; 2, North Platte; 3, Brady; 4, Cozad; 5, Lexington; 6, Overton; 7, Elm Creek; 8, Odessa; 9, Kearney; 10, Gibbon; 11, Wood River; 12, Grand Island; 13, Central City; 14, Clarks; 15, Silver Creek; 16, Duncan; 17, Columbus; 18, Schuyler; 19, North Bend; 20, Fremont; 21, Venice; 22, Ashland; 23, Louisville.

return flows. In the middle of the nineteenth century, travelers along the Oregon Trail were astonished by the width and character of the Platte River in Nebraska. It was unlike any river east of the Mississippi River, and as a result, the pioneers and Army officers commented on it in their journals (Mattes, 1969). By the middle of the twentieth century the river had been greatly modified by hydrologic change, as a result of impoundments and irrigation diversions. A series of maps (1860, 1938, 1957, 1983) prepared by the University of Nebraska's Remote Sensing Applications Laboratory (Peake, *et al.*, 1985) provides a record of this change (see also Eschner *et al.*, 1983). These maps show the active channel and the vegetation type adjacent to the active channel for each US Geological Survey topographic map from about the junction of the North and South Platte Rivers near Brady to just downstream of Grand Island (Figure 7.5). The area of active channel was given for each map so when area was divided by the length of channel an average channel width for each map was obtained for 1860, 1938, 1957, 1983, and 1995 (Figure 7.6).

Human-induced hydrologic changes in the Platte River drainage basin undoubtedly commenced in the late nineteenth century, but continuous hydrologic records only began in the mid-1930s. It was not until the drought years of the 1930s, the completion of Kingsley Dam in 1941 at Keystone (Figure 7.5), and the filling of its reservoir (Lake McConaughy) on the North Platte River that great changes in mean annual discharge, flood peaks, and flow duration were recorded.

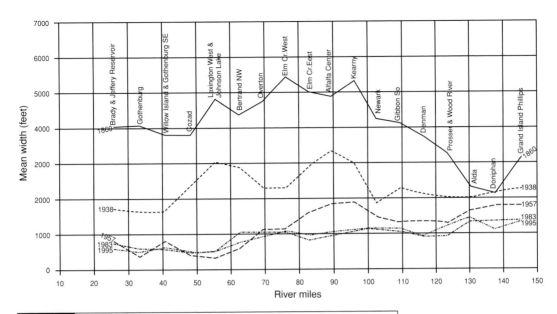

Figure 7.6 Mean width of Platte River in 1860, 1938, 1957, 1983, 1995.

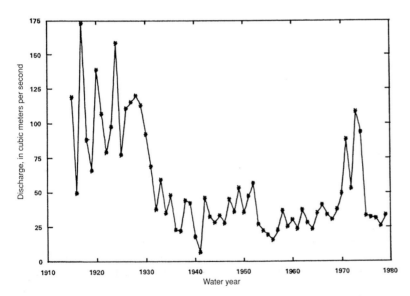

Figure 7.7 Mean annual discharge of Platte River between 1915 and 1979.

During the drought years of the 1930s and following closure of Kingsley Dam, the average annual discharge decreased significantly (Figure 7.7) except for some wet years in the early 1970s. High peak discharges were also less frequent after 1940. These hydrologic changes contributed to the major narrowing of the Platte River (Figure 7.6). Early photographs show that the South Platte, North Platte (Figure 7.8) and Platte River (Figure 7.9) were classic examples of very wide braided streams. The Platte River in 1860 was about a mile (c. 1.6 km) wide.

Figure 7.8 North Platte River in Lincoln County, Nebraska, 1869. (W. H. Jackson photograph courtesy of U.S. Geological Survey.)

Figure 7.9 Platte River 8 km southeast of Grand Island, Nebraska in about 1900. (Photograph courtesy of the Stuhr Museum of the Prairie Pioneer.)

The dramatic changes of the channel width (Figure 7.6) through time require an explanation. One can assume that the river in 1860 was essentially unchanged from natural conditions, although some diversions undoubtedly had commenced. Nevertheless, it should be noted that the Platte River during the 1860 surveys was hydrologically different from the present river. It was intermittent, as described by Ware (1911) in 1863 as follows:

	Period of record	Total number of no-flow days	Mean days of no-flow
Table 7.3 No-flow days at Platte River gages			
Station			
Overton			
Pre-project	1–31–1941	863	78
Post-project	1942–1994	0	0
	1931–1938	657	82
	1939–1957	207	11
	1958–1995	0	0
Odessa			
Pre-project	1939–1941	450	150
Post-project	1942–1957	309	19
Post-project	1957–1991	0	0
	1939–1957	760	40
	1958–1983	3	0
	1984–1995	0	0
Grand Island			
Pre-project	1934–1941	1200	150
Post-project	1942–1978	852	23
Post-project	1978–1995	0	0
	1934–1938	699	140
	1939–1957	1114	59
	1958–1983	253	10
	1984–1995	0	0

From Fort Kearney, for many miles up, there was no water in the river. The water seemed to be in "the under-flow." We not infrequently rode down to the river, and with shovels dug watering-places in the sand of the bed. We always found permanent water within eighteen inches of the top, no matter how dry the sand on top appeared to be. We were told that 75 miles of the river were then dry, and that generally about 125 miles of it were dry in the driest season.

The large number of no-flow days was a characteristic of some reaches of the river until about 1942 (Table 7.3), when the impact of Kingsley Dam and Lake McConaughy on the Platte River became significant. Hydrologic data show that before 1942 (pre-project), the average annual number of no-flow days at Overton (Figure 7.5) was 78, but there were zero no-flow days at this gage after 1941 (post-project). At Odessa there were on average 150 no-flow days a year before 1942, but only 19 no-flow days per year after 1941, and there were zero no-flow days after 1957. At Grand Island, there were an average of 150 no-flow days per year before 1942, but only 23 for the 1942 to 1978 period (Table 7.3). There were zero no-flow days at Grand Island after 1978. Clearly, the hydrologic character of the river had changed from intermittent to perennial, which undoubtedly had a major effect on channel width. The large number of no-flow days in the 1930s also reflected the drought conditions of that decade.

Johnson (1994, p. 77) attributed width reduction during the drought years to low flows, which exposed large areas of the channel bed to colonization by vegetation (woodland expansion). The low flows maintained the water table, whereas a long series of no-flow days caused lowering of the water table and a high mortality of seedlings. Therefore, conversion of the Platte River between Overton to Grand Island from an intermittent river with many no-flow days to a perennial river allowed colonization of the exposed channel bed by vegetation (woodland expansion) and major narrowing of the channel (Figure 7.6).

Johnson's (1994, 1998) conclusions and the recognition of the nature of the hydrologic changes in the Platte River permit the development of explanations of the width changes between 1860 and 1995 (Figure 7.6). The marked decrease of width between 1860 and 1938 can logically be attributed to the effects of diversions, large dam construction and the drought years of the 1930s. The limited hydrologic data for this period show that the number of no-flow days at Overton and Grand Island were numerous (Table 7.3).

For the period 1939 through 1957, an additional significant decrease of width occurred between Brady and Grand Island (Figure 7.6). The average annual number of no-flow days for this period at Overton was 11 per year, whereas at Grand Island the average was 59 per year (Table 7.3). At Overton there were zero no-flow days after 1941, and the channel adjusted to discharges released from Lake McConaughy and irrigation return flow. By 1957, width upstream of the Elm Creek West quadrangle (Figure 7.6) appears to have stabilized, and there were only minor width changes between 1957 and 1995. However, downstream of the Elm Creek West quadrangle, the decrease of width continued through 1983. It was not until 1978 that there were zero no-flow days at Odessa and Grand Island. The absence of no-flow days permitted adjustment of this part of the channel to a relatively stable condition during the period 1984–95 (Figure 7.6).

In summary, at Overton, Odessa, and Grand Island, river width decreased as irrigation return flows eliminated no-flow days between Overton and Grand Island. The establishment of perennial flow and a raised water table promoted vegetation establishment on the floodplain and in the channel. A similar conclusion was reached by (Nadler and Schumm, 1981) for the South Platte River. In contrast, a series of no-flow days in the wide sandy channel created a harsh environment for plant growth. The bare sand surface and the decline of the water table prevented survival of plants that were established in the channel during previous wetter months.

During low flow, the wide pre-1938 river made large areas available for colonization by plants (Johnson, 1994, p. 77), but the probability of mortality later in the year was high, especially when there were a number of no-flow days. The conversion of the Overton to Grand Island channel from intermittent to perennial undoubtedly maintained a high water table and favorable conditions for colonization and survival of vegetation. Therefore, the change of channel width downstream of Overton after 1941 was due to the modification of flow characteristics by the construction and operation of Kingsley Dam and Lake McConaughy (Kircher and Karlinger, 1983; Simons, and Simons, 1994).

Nature of change

The trend of average channel width downstream of Brady (Figure 7.6) in 1860 is unusual. Average width generally increased in a downstream direction from the Brady quadrangle to the Kearny quadrangle, as expected. However, between the Kearny and Doniphan quadrangles, average width decreased dramatically. In this reach, the river was not braided; but rather it was anastomosing. That is, the single braided channel (Figure 7.10a, b) became a multiple channel complex (Figure 7.11a). The total width of the multiple anastomosing channels was less than the width of the upstream braided channel. If the same volume of water moved through the braided channel at Kearny, as through the anastomosing channels down-stream, then each anabranch must have been deeper than the braided channel. Nanson and Huang (1999) conclude that an anastomosing channel will be narrower than a single channel. A possible explanation for this pattern change is that the gradient was about 5 percent less in the anastomosing reach.

The average gradient between Cozad and Kearny is 0.00130, whereas the average gradient between Kearny and Grand Island is 0.00124. In addition, the valley widens at Kearny and the contours on topographic maps are no longer deflected upstream, as they cross the river. This suggests that the anastomosing reach occurs where flood waters are likely to spread across the valley and form multiple channels. In the braided reaches, even during high water, the banks were not overtopped according to accounts of the early travelers (Mattes, 1969, pp. 163–164, 240), but near Kearny the river had the appearance of flowing at the level of the valley floor (Mattes, 1969, pp. 163, 240). Near Grand Island in 1860, the river reverted to a braided pattern, which is the present condition.

The two different channel patterns of the Platte River in 1860 responded differently to the hydrologic changes that caused width reductions between 1860 and 1995. The Platte River between Brady and Grand Island provides an excellent example of river variability in location and through time. For example, the braided reach of 1860 (Figure 7.10a) by 1938 contained many more vegetated islands, and it had become an island-braided river (Figure 7.10b). However, the river undoubtedly contained vegetated islands in 1860, but they were ignored by the early surveyors. Nevertheless, these islands coalesced and increased in size, the single-channel braided stream became a smaller multiple-channel anastomosing

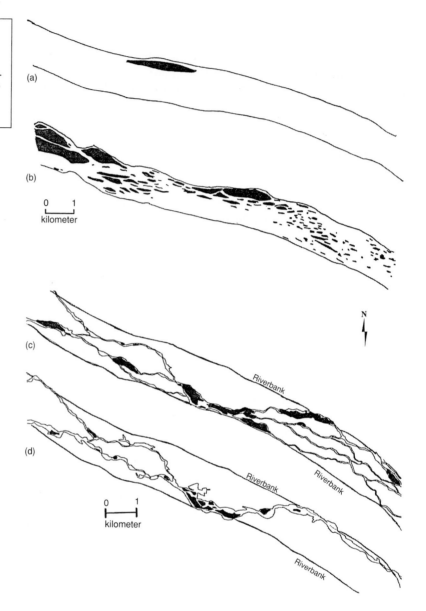

Figure 7.10 Platte River on Cozad quadrangle (Figure 7.6) in (a) 1860, (b) 1938, (c) 1957, and (d) 1983 (from Peake et al., 1985). Dark areas are islands. 1860 river bank indicates former channel width.

river (Figure 7.10c), which with time and abandonment of secondary channels, became a much narrower single braided channel (Figure 7.10d). In contrast, in the anastomosing reach (Figure 7.11a) near Newark just east of Kearney (Figure 7.7), two secondary anabranches, which were narrower and shallower channels, were abandoned between 1860 and 1938. Two channels remained, the northern anastomosing channel much reduced in size, and the southern, apparently dominant, braided channel (Figure 7.11b). By 1957 the northern channel was becoming a narrow single channel and the southern channel remained essentially as it was in 1860 (Figure 7.11c). In 1983 the northern channel was still approaching a

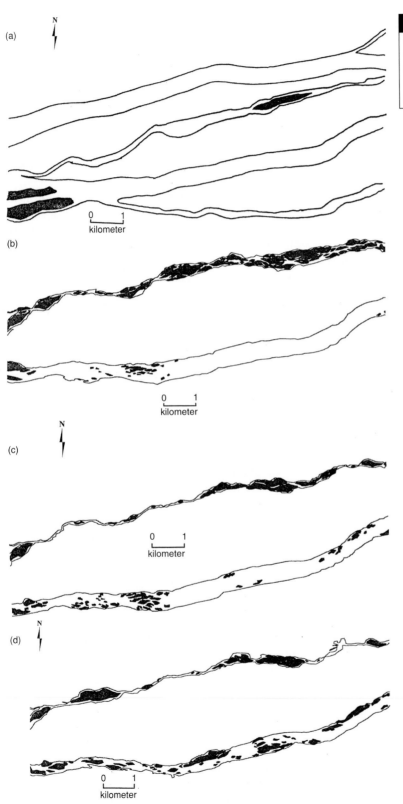

(a)

(b)

(c)

(d)

Figure 7.11 Platte River anastomosing channels on Newark quadrangle (Figure 7.6). Dark areas are islands (from Peake *et al.*, 1985), (a) 1860, (b) 1938, (c) 1957, and (d) 1983.

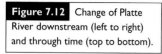

Figure 7.12 Change of Platte River downstream (left to right) and through time (top to bottom).

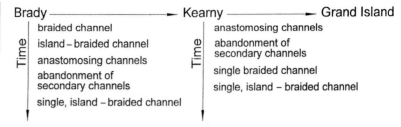

single-channel morphology, but the southern channel was island braided (Figure 7.11d).

Although the Platte River became narrower between 1860 and 1995 (Figure 7.6), the adjustment differed between the braided (upstream of Kearny) and anastomosing reaches (Kearny to Grand Island). Figure 7.12 presents an idealized evolutionary sequence of channel changes through time for both types of channel, although neither type has achieved the final stage of a single channel.

The Platte River in Nebraska provides a good example of the response of braided and anastomosing channels to hydrologic change. The braided channel narrowed significantly by island formation and the development of small side channels (anastomosing), which were then abandoned. The result was a narrower braided channel. The anastomosing reach was converted to a narrow island-braided channel by abandonment of secondary channels and concentration of flow in one channel. Reduced mean annual discharge, reduced peak discharge, and the development of perennial flow all contributed to the metamorphosis of the Platte River.

In terms of downstream variability, hydrologic and climatic variability can strongly influence channel morphology. A tributary changing sediment loads and flow characteristics, although a local control (Chapter 10), can transform a river from braided to meandering and vice versa.

Although largely the effect of human activity, the changes of the Platte River indicate how a channel can adjust if the water-table rises or falls naturally. It is likely that the Cimarron River responded to drought and a falling water table which caused loss of riparian vegetation. Large floods then easily attacked the banks. With the return of more humid conditions and a rising water table, vegetation colonized the higher parts of the channel and the channel was significantly narrowed like the Platte River.

Mean annual discharge, flood peaks, and even no-flow days have major impacts on channel morphology. Even a reversal of the hydraulic geometry relations occurs when discharge decreases downstream.

Chapter 8

Humans

According to Dynesius and Nilsson (1994), 77 percent of the 139 largest river systems in North America (north of Mexico), in Europe and in the republics of the former Soviet Union are affected by dams, reservoir operation, interbasin diversions, and irrigation. It is obvious that humans have impacted rivers since the beginning of civilization in the valleys of the Yellow, Indus, Tigris-Euphrates, Ganges, and Nile rivers (Figure 1.2). In fact, the earliest dam was constructed on a tributary of the Nile south of Cairo about 5000 years ago (Schnitter, 1994).

Graf (2001) calculates that 79 percent of American rivers are affected by humans (Table 8.1), and he developed a scale of naturalness that is illustrated by five American rivers (Table 8.2). Much of the fragmentation of rivers in the United States is due to 80 000 dams that have been constructed on these rivers.

Wohl (2001) states that when she moved to Colorado in 1989 she "was impressed by the sparkling water of the mountain rivers, . . . and assumed that these were natural fully functional rivers." However, when she "began to read historical accounts of the Colorado Front range and to examine the streams more closely" she "realized how dramatically they had been altered." She began to "think of them as virtual rivers, which had the appearance of natural rivers but had lost much of a natural river's ecosystem functions."

Human impacts on rivers are the results of a variety of activities, perhaps the greatest impact is the change of flow and sediment load downstream of dams, but placer mining, gravel mining, dredging, levee construction, ripraping, channelization, etc., all have their effects. As noted earlier, river response to these impacts will vary depending upon sediment characteristics, climate, and the type of river, which results in uncertainty and unpredictability concerning the effect of controls (Phillips, 2002). Some of the above listed impacts have only a local effect, but channelization and the modification of discharge and sediment load by dams can transform long reaches of a channel.

Channelization usually involves the straightening of a sinuous channel to reduce overbank flooding. The result is a very unstable channel that deepens and widens and eventually achieves a degree

Table 8.1 Estimated length of American rivers

Physical Condition	Length (km)	Length (mi)	Percent of total
Affected by human activities	4 022 400	2 514 000	78.9
Drowned by reservoirs	960 000	600 000	18.8
Unaffected by human activities	40 000	64 000	2.0
In the Wild and Scenic Rivers System	17 304	10 815	0.3
TOTAL	5 120 000	3 200 000	100.00

(From Graf, 2001)

of stability (Figure 3.1). These artificially incised channels mirror the natural changes of arroyos in southwestern USA (Schumm *et al.*, 1984; Gellis *et al.*, 1991; Darby and Simon, 1999; Elliott *et al.*, 1999; Hupp, 1999). The channel evolution requires about 40 years in humid Mississippi (Figure 3.1), but about 100 years in dry New Mexico and Arizona (Figure 3.2).

Dams, which modify both sediment loads and discharge, usually cause degradation below the dam and channel narrowing (Petts, 1979; Williams and Wolman, 1984; Collier *et al.*, 2000), but armoring can have a significant effect (Chapter 19). The changes of the Platte River, as described in Chapter 7, provide an excellent example of major channel response and colonization of portions of the channel by vegetation (Johnson, 1998). Friedman *et al.* (1998) studied the effects of 35 dams on rivers of the Great Plains and the central USA. They found that braided streams narrowed and the rate of migration of meandering streams decreased. Invasion of woody vegetation, as along the Platte River, caused narrowing of the braided streams. The meandering streams suffered less vegetation encroachment, but bank erosion increased, and the channels deepened.

If a single example of the impact of humans on a great river system is needed, the Mississippi, Ohio, and Missouri River system (Figure 1.3) provides an impressive example, as these rivers were modified for flood control and navigation. A comparison of the suspended sediment discharge from the Mississippi River basin (Meade, 1996) for pre-settlement (about 1700) and present conditions reveals major changes (Figure 8.1). Sediment delivery from the Ohio River has increased. Although a series of locks and dams maintain flow necessary for navigation, there are no sediment trapping dams on the Ohio River channel and deforestation and agricultural activities have increased sediment loads in tributaries and in the Ohio River. The upper Mississippi River (Figure 1.3) above the junction of the Missouri apparently has not increased sediment loads significantly in spite of increased agriculture and deforestation (Figure 8.1). A series of

Table 8.2 Scale of naturalness for the geomorphology of river channels

Components of channel physical integrity	Geomorphological naturalness				
	Completely natural	Partly modified	Substantially modified	Mostly modified	Completely artificial
Channel pattern	Pretechnological pattern, often meandering single thread or complex braided	Minor portions of the pattern altered by engineering works	About half of the channel pattern is engineered	Only remnants of pretechnological pattern remain	Completely engineered channel pattern, usually straightened and single thread
Channel cross-section	Pretechnological cross-section, often highly complex	Minor portions of the cross-section altered by human activities	About half of the cross-section is altered by human actions	Only remnants of the pretechnological cross-section remain	Designed, completely engineered cross-section, usually highly simplified
Minor landforms (bars, islands, pools, riffles), functional surfaces and materials	Pretechnological sizes and distribution of minor landforms, often highly complex	Minor changes in size or distribution of minor landforms as a result of human activities	About half of the minor landforms of the reach altered by human activities	Only remnants of the pretechnological minor landforms remain	Minor landforms in the channel eliminated by engineering, or artificial forms included by design
Descriptive notes	River undisturbed by technological activities, could be a "wild" river under the Wild and Scenic Rivers Act	River retaining much of its pretechnological characteristics, but with some modifications and/or impacts, often from altered flows of water or sediment	Combined "natural" and artificial river, with obvious intentional modifications and/or unintended human impacts	River with major extensive modifications and/or impacts	River as a product of design and engineering
Example	Middle Fork of the Salmon River, Idaho	Lower St. Croix River, Wisconsin and Minnesota	Concord River, Massachusetts	Lower Colorado River, Arizona and California	Los Angeles River in downtown Los Angeles, California

Source: Simplified and modified from Graf (2001)

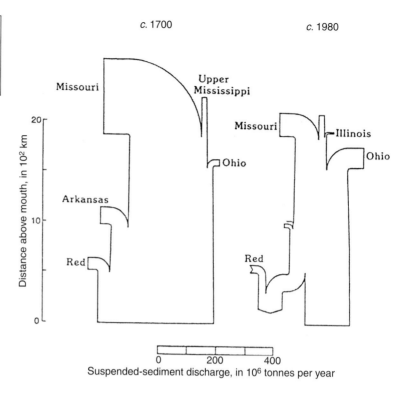

Figure 8.1 Long-term average discharges of suspended sediment in the lower Mississippi River in c. 1700 and c. 1980 (from Meade, 1996).

25 navigation locks and dams formed large pools (Figure 8.2) that may be trapping the sediment upstream of the locks and dams. However, the western tributaries, Missouri, Arkansas, and Red Rivers all show significant decreased sediment loads. Five large flood control dams on the Missouri trap much of its sediment load, and dams trap sediment in the Arkansas and Red River drainage systems.

The dramatic decrease of sediment delivery to the main channel of the Middle Mississippi River (Missouri River junction to Ohio River junction) and to the Lower Mississippi River (Ohio River junction to the delta) would be expected to have a major impact on the Mississippi River proper, but the channel has been so significantly modified by levee construction, cutoffs, and navigation channel improvements (dredging, dikes, riprap) that detection of the upstream impacts is difficult.

The Mississippi River can be divided into three major reaches as follows: (1) Upper Mississippi River upstream of the Missouri River junction; (2) Middle Mississippi River between the junction of the Missouri River and the Ohio River; and (3) Lower Mississippi River, downstream of the Ohio River junction to the head of the delta at about Baton Rouge, Louisiana (Figure 1.3), and each will be discussed briefly below.

Upper Mississippi River

The Upper Mississippi River in its natural state was island braided. The islands were vegetated and stable. The geomorphic changes of the

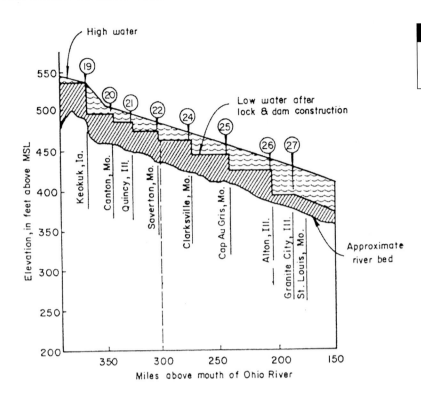

Figure 8.2 Longitudinal profile of a reach of the Upper Mississippi River showing effects of locks and dams 19–27. MSL, mean sea level.

Upper Mississippi River through historic time have been primarily in response to the influence of humans. Navigation improvements have been under way since 1824. The first attempt to improve conditions was to remove snags hazardous to navigation. Later dikes were constructed to confine the low flows to a narrow channel, thus increasing the depth of flow. At the same time, revetment was placed along caving banklines to hold the channel alignment. In the 1930s the navigation channel depth was increased to 9 feet (c. 3 m) by constructing 26 locks and dams. Since that time, the navigation channel has been maintained by these structures and supplemental dredging. In addition, there was excavation of rock in several reaches of rapids; and closing off of meandering sloughs and backwaters to confine flows to the main channel and thus assure more adequate depths for navigation during times of low-water flow. As an extreme example of human impacts in the 19th century, the great number of saw mills and debris from their operation produced sawdust bars in the river upon which steamboats often went aground (Merritt, 1984, p. 15).

Viewed in the longitudinal profile, the locks and dams on the Upper Mississippi River 9-foot (c. 3-m) channel project form a series of steps in a river stairway (Figure 8.2). They are not storage dams like the high dams on the Missouri River; instead they regulate river flows to maintain the minimum 9-foot (c. 3-m) depth required for navigation.

All the activities, construction of dikes and locks and dams, operation of locks and dams, and dredging, have changed the river. To

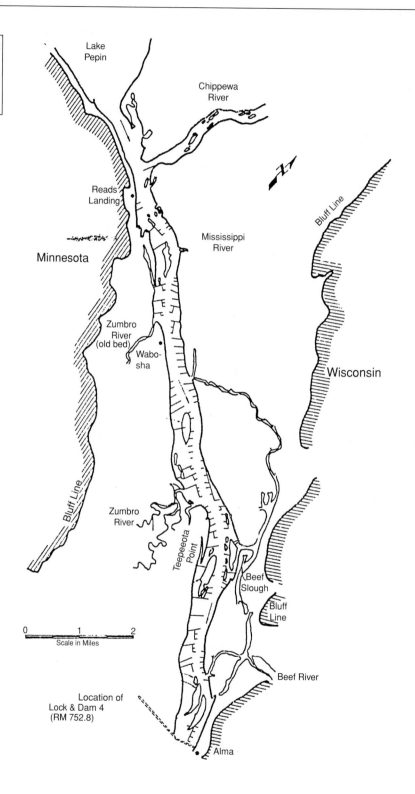

Figure 8.3 Dikes in Pool 4, Upper Mississippi River, downstream of Chippewa River junction (Figure 1.3), before construction of lock and dam 4.

Lake Pepin

Chippewa River

Reads Landing

Mississippi River

Minnesota

Zumbro River (old bed)

Wabo-sha

Wisconsin

Bluff Line

Bluff Line

Zumbro River

Teepeeota Point

Beef Slough

Bluff Line

Beef River

0   1   2
Scale in Miles

Location of Lock & Dam 4 (RM 752.8)

Alma

obtain an understanding of the response of the river system to these activities, the past and present character of the river upstream of Lock and Dam 4 and downstream of the Chippwa River is reviewed. The river reach known as Pool 4 is generally typical of most of the downstream pools (Simons *et al.*, 1976). Just upstream of Lock and Dam 4, the change of water levels submerged the dikes constructed during the earlier channel projects (Figure 8.3), and the initial operation of Lock and Dam 4 in 1935 caused inundation of a large floodplain area upstream of the structure as shown in the 1973 map (Figure 8.4). Hence the width of the river in the lower one-third of Pool 4 increased (Table 8.3) whereas the upper one-third narrowed, as a better defined low-water channel developed.

Therefore, the modern upper Mississippi River consists of a series of reaches each of which is confined between locks and dams. The general character of each reach is a narrow upstream segment, a middle transitional segment, and a wider expanse of water in a pool upstream of the lock and dam.

Middle Mississippi River

The Mississippi River changes its morphology dramatically at the confluence of the Missouri River and the type of river stabilization measures also changes. As in the upper river, the Corps of Engineers was authorized by Congress to obtain and maintain a 9-foot (*c.* 3-m) deep 300-foot (*c.* 91-m) wide navigation channel through the Middle Mississippi River (Figure 1.3). This was achieved by the construction of dike fields (Figure 8.5) to reduce channel width.

In 1973, almost the entire river from the mouth of the Missouri River north of St. Louis was lined with mainline levees on one bank or the other. One hundred twenty-two miles (*c.* 195 km) of bankline revetment prevent riverbank erosion and over 800 dikes project out from the river banks into the river channel.

The Middle Mississippi River is remarkably straight (Figure 8.5), unlike the island braided Upper Mississippi and the meandering Lower Mississippi. In addition, it is a relatively stable river. The objectives of flood protection and year-round river navigation have been met to a great extent on the Middle Mississippi River. However, the developments for flood protection and river navigation have produced a new river morphology and a different river behavior.

The Middle Mississippi River has been deepened for navigation by decreasing channel width with rock and pile dikes. An example of the change of cross-sectional geometry is shown in Figure 8.6. In 1837 the river at St. Louis was 3700 feet (*c.* 1.1 km) wide, and it had an average depth of 30 feet (*c.* 9m) at bankfull stage. The dike construction which was started at St. Louis in 1838 and completed before 1888 permanently decreased the width to 2100 feet (*c.* 640 m). In 1973 the average bankfull depth was about 45 feet (*c.* 14 m). The cross-sectional

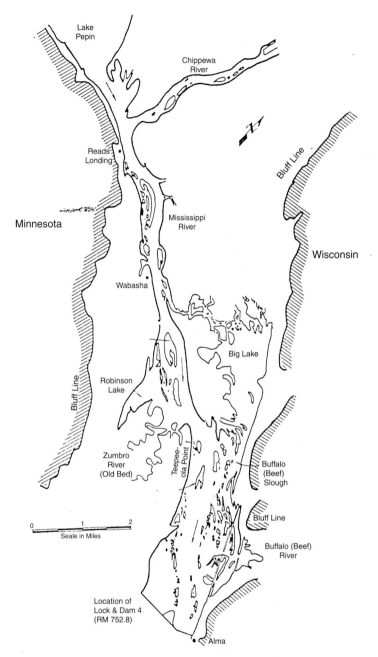

Figure 8.4 Character of Pool 4 Upper Mississippi River after construction of Lock and Dam 4.

area at the bankfull stage was approximately 80 000 square feet ($c.$ 24 384 m^2) in 1973, whereas it was 120 000 square feet ($c.$ 36 576 m^2) in 1837, and width–depth ratio had decreased from 123 to 47. The narrowing of the channel at St. Louis reduced the bankfull channel area by about one-third. Narrowing the river at St. Louis caused a general degradation of the bed, and the bed was on the average 8 feet ($c.$ 2.4 m) lower after contraction. Riverbed degradation has occurred along the Middle Mississippi River wherever the channel has been narrowed.

Table 8.3 | Average river surface widths of Pool 4 below the mouth of the Chippewa River

Location	Surface width (Feet)			
	1850	1897	1929	1973
Upper one-third	2440	2410	1890	1740
Middle one-third	2430	2410	2300	3200
Lower one-third	2590	2550	2550	5560

Note: 1 ft, c. 0.3048 m

Figure 8.5 Dikes in Middle Mississippi River.

Discharge in the Middle Mississippi River has been measured at St. Louis intermittently from 1843–61, and continuously since 1861. St. Louis is below the confluence of Missouri and Upper Mississippi Rivers. The maximum flood discharge, 1 300 000 cfs (cubic feet/second) occurred at St. Louis in 1844. The Missouri River contributed 900 000 cfs of this, which was the record flood for this river.

The construction of levees along the floodplain affected the natural flows of the Middle Mississippi. The floodplain is a storage area for flood waters when the river rises above bankfull stage. Levees increase the flow height for discharges greater than bankfull. The increase in stage results from the decrease in floodplain storage.

The effects of 100 years of development in the Middle Mississippi River on river stage were clearly illustrated during the 1973 flood. The 1973 peak flood discharge at St. Louis was 852 000 cfs, which resulted in a maximum record high-water stage of 43.3 feet (c. 13 m). The 1844 record discharge of 1 300 000 cfs would now pass St. Louis with an estimated stage of 52.0 (c. 16 m) feet instead of 41.3 feet (c. 12.5 m) in 1844, an 11-foot (c. 3.5 m) increase.

The largest flood discharges at St. Louis are listed in Table 8.4; the 1973 flood ranks no. 10. The record high stages are listed in Table 8.5;

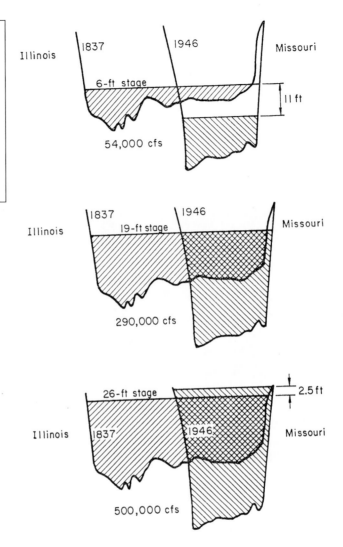

Figure 8.6 Changes in cross-section area of Mississippi River at St. Louis between 1837 and 1946 for discharges of 54,000 cfs, 290,000 cfs, and 500,000 cfs. At discharges below 290,000 cfs the water level will be lower, whereas at discharges above 290,000 cfs the water level will be higher (from Stevens et al., 1975a). ft, feet; cfs, cubic feet per second (0.0283 m³/s).

the 1973 flood-stage ranks no. 1. The narrowing and deepening of the channel, which improved navigation, had the unintended consequence of raising flood stage (Belt, 1975; Stevens *et al.*, 1975a; Criss, 2001). For example, the large 1973 flood ranked tenth according to discharge (Table 8.4), but it ranked first according to stage (Table 8.5). The huge 1844 flood was demoted to second place. The explanation, of course, was that the 1844 flood spread across the floodplain, whereas the 1973 flood was confined between levees (Figure 8.7).

The construction of dikes in the Middle Mississippi River caused channel narrowing as sediment was deposited in the dike fields (Figure 8.5). Eventually the deposited sediment was vegetated, and an island formed with a narrow side-channel separating the island from the floodplain (Figure 8.5). The side channel is a favorable habitat for fish and wildfowl, but eventually it fills with sediment, and the island will join the floodplain.

Table 8.4	Top 10 flood discharges at St. Louis	
Rank	Peak discharges, in cubic feet per second	Year
1	1 300 000	1844
2	1 054 000	1858
3	1 050 000	1855
4	1 040 000	1903
5	1 022 000	1851
6	926 000	1892
7	889 000	1927
8	863 000	1883
9	861 000	1909
10	855 000	1973

Note: 1 cfs, *c.* 0.0283 m^3/s

Table 8.5	Top 10 flood stages at St. Louis	
Rank	Maximum stage, in feet	Year
1	43.3	1973
2	41.3	1844
3	40.2	1947
4	40.2	1951
5	39.0	1944
6	38.9	1943
7	38.0	1903
8	37.2	1858
9	37.1	1855
10	36.6	1851

Note: 1 ft, *c.* 0.3048 m

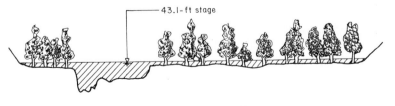

The 1844 river cross-section

Figure 8.7 Effect of levees and narrowing of Mississippi River on water level at a discharge of 1,300,000 cfs, the maximum flood of 1844 (from Stevens *et al.*, 1975a). cfs, cubic feet per second (0.283 m^3/s).

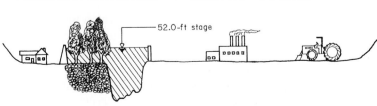

The 1973 river cross-section

Table 8.6	Man-made neck cutoffs, 1929–1942				
		River mile location on 1975 Maps	Year opened	Bendway miles	Cutoff distance miles	Distance river shortened miles

	River mile location on 1975 Maps	Year opened	Bendway miles	Cutoff distance miles	Distance river shortened miles
678	Hardin	1942	18.8	1.9	16.9
628	Jackson	1941	11.1	2.4	8.7
625	Sunflower	1942	12.9	2.5	10.4
575	Caulk	1937	17.2	2.0	15.2
549	Ashbrook	1935	13.3	1.9	11.4
541	Tarpley	1935	12.2	3.6	8.6
589	Leland[a]	1933	11.2	1.4	9.8
514	Worthington	1933	8.1	3.8	4.3
504	Sarah	1936	8.5	3.2	5.3
463	Willow	1934	12.4	4.7	7.7
448	Marshall	1934	7.3	3.1	4.2
424	Diamond	1933	14.6	2.6	12.0
408	Yucatan[a]	1929	12.2	2.6	9.6
388	Rodney	1936	10.0	4.1	5.9
366	Giles	1933	14.0	2.9	11.1
343	Glasscock	1933	15.6	4.8	10.8
TOTALS			199.4	47.5	151.9

[a] Natural cutoffs.

1 river mile, c. 1.61 km.

Lower Mississippi River

The Lower Mississippi River (Figure 1.3) extends from Cairo, Illinois and the mouth of the Ohio River to Baton Rouge, Louisiana about 230 miles (c. 370 km) upstream from the Gulf of Mexico (Winkley, 1977). This segment of the Mississippi River has been subjected to the full force of river control. Winkley (1977, 1994) provides a summary of this work. Levee building commenced in about 1719 and the length and height of levees increased with time cutting off access to flood basins, which confined about 600 000 cfs (c. 16 980 m³/s) to the river during floods (Winkley, 1977). Bank protection also increased and dredging was initiated in order to maintain navigation. Nevertheless, perhaps the greatest impact that transformed portions of the river from highly sinuous to straight was the meander cutoff program of the 1930s and 1940s which shortened the river 152 miles (c. 245 km) and another 55 miles (c. 89 km) by chute development. The chutes were the result of the neck cutoffs which increased flow velocity causing incision upstream and deposition downstream.

Table 8.6 is a list of cutoffs shown on Figure 8.8. The cutoff program was initiated following the great 1927 flood, which devastated the valley. Between 250 and 500 people were killed and 650 000 were homeless, 165 000 livestock drowned, and 26 000 square miles

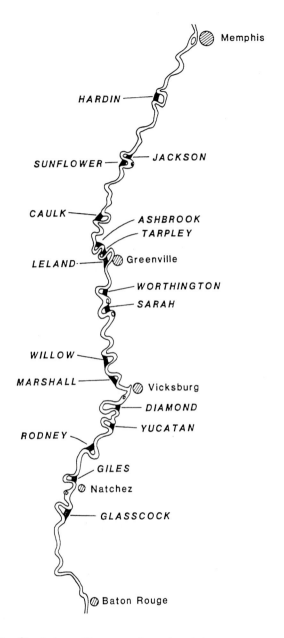

(c. 41.6 km²) of the valley were inundated (Barry, 1997). In order to speed the flow and reduce flood peaks, cutoffs were made. Indeed, flood peaks were reduced, but channel stability decreased.

The most dramatic shortening of the river was the straightening of the Greenville Bends (Figure 8.8) by the Ashbrook, Tarpley, and Leland cutoffs in 1933 and 1935 (Table 8.6). In this reach, the original river length was 37 miles (c. 59 km). The reach was shortened 29.8 miles (c. 48 km) or 81 percent of its original length, which increased the gradient through the reach about fivefold. The increased velocity caused severe bank erosion, and the river in 1975 was wider than the 1933 river. According to Winkley (1977), in spite of millions of dollars

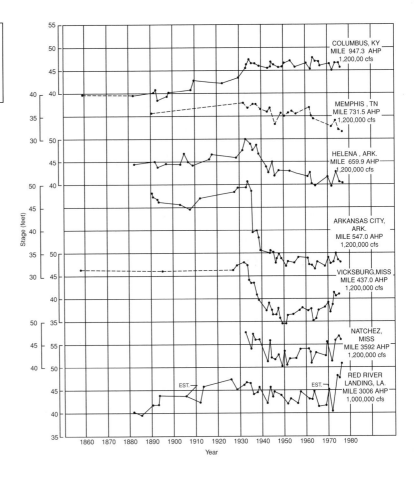

Figure 8.9 Water level changes after 1930 for the Mississippi River showing effect of cutoff program. AHP, head of passes; cfs, cubic feet per second (0.0283 m³/s).

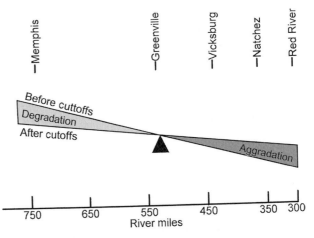

Figure 8.10 General effect of cutoffs on Mississippi River showing degradation upstream of Greenville and aggradation downstream. 1 river mile, 1.61 km. From Winkley (1977).

spent in dike and revetment construction and dredging, the river in 1977 was still attempting to return to its natural meandering state rather than its braided condition.

The cutoffs were concentrated in a 340 mile (*c.* 547 km) reach of the river, 55 miles (*c.* 89 km) downstream of Memphis to 23 miles (*c.* 37 km)

Part III

Fixed local controls

Local controls can be fixed in position (Figure 1.2) by bedrock, resistant alluvium, tributaries and tectonic features.

Chapter 9

Bedrock: alluvium

Bedrock, resistant alluvium and gravel armor constrain channel behavior and morphology (Figure 1.2). Rivers that are confined between bedrock canyon walls are obviously dominated by bedrock. However, many rivers are only partly affected by bedrock, and they vary greatly in their morphology and ability to adjust in a downstream direction. For example, the Snake River in Idaho (Osterkamp *et al.*, 2001), the Middle Fork of the John Day River in Oregon (McDowell, 2001), and the Sabie River in South Africa (Heritage *et al.*, 2001) are of this type. Mapping of 25 km of the Sabie River reveals that 5 km is bedrock anastomosing, 6.5 km is pool-rapid, 8 km is mixed bedrock and alluvial anastomosing, 1 km is single thread, and 4.5 km is braided, a very mixed bag.

For eight major rivers of the Northern Rocky Mountains in Idaho, it is obvious that the rivers are strongly influenced by the presence of bedrock, the nature or type of bedrock (resistant, soft, jointed, faulted), the presence of resistant alluvium in terraces and alluvial fans, colluvium (debris flows, rock falls), and glacial outwash. Reaches of these rivers were mapped in order to determine how much of the channels were controlled by bedrock, alluvial fans, and terraces, and indeed, a considerable portion of the bed and banks of these rivers is controlled by these variables (Table 9.1).

In fact, the length of channel with no impact of bedrock, alluvial fans, or old alluvium comprises a small percentage of total channel length. Figure 9.1 illustrates how alluvial fans, rock falls, and jointing affect channel width and flow direction. Bedrock resistance also determines valley width (Wohl and Ikeda, 1998, Warner, 1983), channel width, and the ability of a channel to shift laterally. Obviously, none of the above described channels are of the regime type. For example, within a 3 km reach of the Idaho rivers, channel width varies from 12 to 104 m and, of course, the longitudinal profiles are irregular (Figure 9.2). Figure 9.3 shows how the gradient of alluvial reaches can be controlled by discontinuous bedrock outcrops. If gradient is measured on alluvium between locations 1 and 2 in each reach, the results will be different, and the gradient of one reach will not be representative of adjacent reaches.

Table 9.1 | Mapped length of rivers and length of geologic-geomorphic controls

River	Total length of reach (km)	Bedrock (m)	Fans (both banks) (m)	Terraces (both banks) (m)	No control (m)	Length of channel with no control (%)
Boise River	16.6	12 505	3202	17 995	0	0
South Fork Payette	39.9	27 450	9455	58 255	763	2
South Fork Salmon	29.0	1342	3965	17 690	2440	8
Lochsa River	107.4	61 000	20 130	66 185	5795	5
Selway River	29.9	1678	5795	14 335	1220	4
North Fork Clearwater	41.7	35 380	13 725	10 370	0	0
East Fork, South Fork Salmon	22.5	17 385	2013	4270	3508	16
Middle Fork Clearwater	36.2	15 555	9150	31 110	2135	6

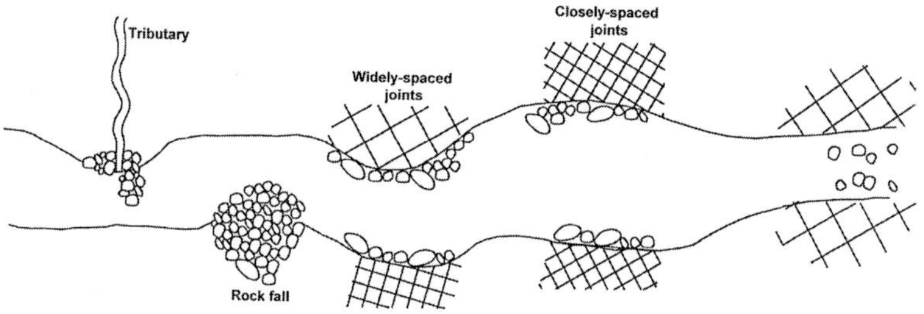

Figure 9.1 Sketch showing types of geologic–geomorphic controls on width of North Fork Clearwater River, Idaho (from Schumm, 1994).

Even when bedrock is relatively weak, the impact on even very large rivers can be significant. For example, the Mississippi River between the Ohio River junction and Old River north of Baton Rouge can be divided into 24 reaches (Figure 9.4) based upon valley slope, sinuosity and rate of change (Schumm *et al.*, 1994). The boundary between reaches in 21 cases involves Tertiary-age "bedrock" (Reaches 1, 2, 3, 4, 8, 9, 10–24) and Pleistocene-age gravel. In 12 cases, uplift and faulting is associated with the more resistant sediments (Reaches 3–7, 9, 10, 12, 17, 18, 21) and in five cases there are also tributary influences (Reaches 13, 14, 16, 17, 24).

Floodplain sediments and bank sediments are controls on channel behavior (Fisk, 1944, 1947; Schumm and Thorne, 1989). It is obvious that sandy banks will erode more rapidly than clayey banks. Maps of the Mississippi River show how the impact of isolated clay deposits,

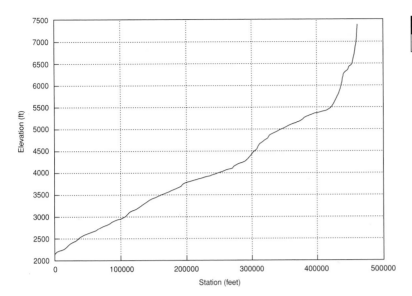

Figure 9.2 Longitudinal profile of South Fork Salmon River, Idaho.

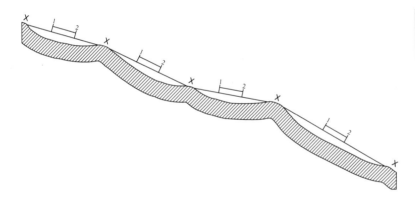

Figure 9.3 Bedrock outcrops (X) in river bed determine gradient of reaches.

clay plugs (Figure 9.5), influence meander migration and meander shape (Figure 9.6).

When an alluvial meander encounters resistant sediments or bedrock, the downstream limb of the meander will be fixed in position and the upstream limb will continue to migrate, thereby deforming the meander (Figure 9.6a). A meander increasing in amplitude will develop a flat top, or a multiple bend when it encounters resistant material (Figure 9.6b, c). Finally, a sequence of meanders may be deformed as they shift downvalley toward resistant material (Figure 9.6d).

A river entering a region of resistant floodplain sediments such as backswamp deposits in the Mississippi Valley will develop characteristic stable bends of relatively high amplitude. For example, the farthest downstream bends of the Mississippi River in fine back-swamp alluvium are very different from those upstream that are formed in sandy alluvium, and the rate of change is much higher upstream (Figure 9.7).

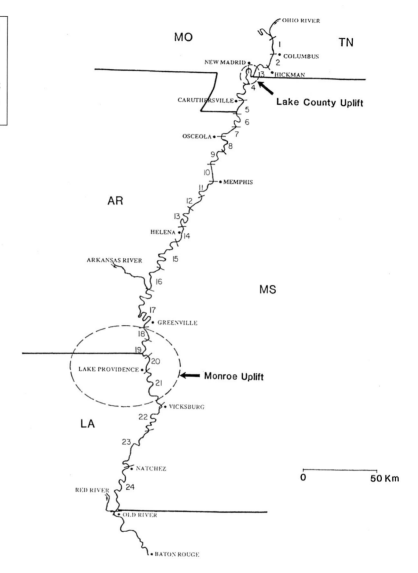

Figure 9.4 Twenty-four reaches of the lower Mississippi River (from Schumm et al., 1994). Also shown is location of Lake County and Monroe Uplifts. MO, Missouri; TN, Tennessee; AR, Arkansas; MS, Mississippi; LA, Louisiana.

Examination of maps of the Mississippi River reveal that most of the meanders are deformed as a result of the effect of abandoned-channel fills of cohesive sediments. These "clay plugs" are very common, and when the river encounters them, the results are irregular bank lines, meanders with flattened crests and compressed and irregular meanders. In addition, the cross-section is significantly altered. When the river impinges on a clay plug, the channel narrows and deepens with significant effects on flow characteristics and bed forms (Fisk, 1947, Krinitsky, 1965).

In an unpublished report, Harold Fisk (1943) cites several examples of the effect of clay plugs on bend morphology and bank erosion:

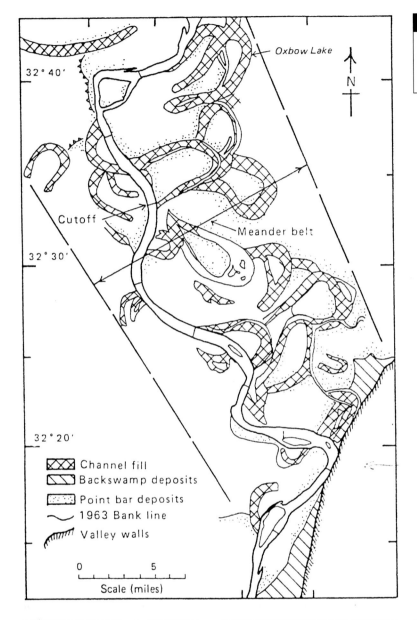

Oxbow Lake

32° 40'

N

Cutoff

Meander belt

32° 30'

32° 20'

Channel fill
Backswamp deposits
Point bar deposits
1963 Bank line
Valley walls

0   5

Scale (miles)

Figure 9.5 Effect of clay channel fills on Mississippi River north of Vicksburg, Mississippi. The river is relatively straight as it is confined by resistant sediments.

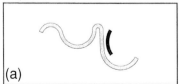

(a)

(b)

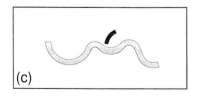

(c)

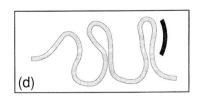

(d)

Figure 9.6 Effect of obstructions (bedrock, resistant alluvium) on meander patterns.

Figure 9.7 Character of meanders, Mississippi River. (a) Upstream in sandy sediments; (b) downstream in clayey sediments.

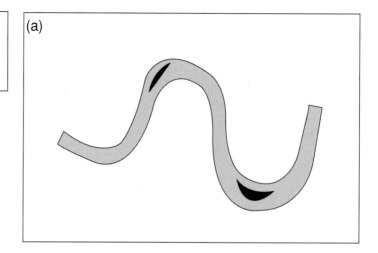

(a)

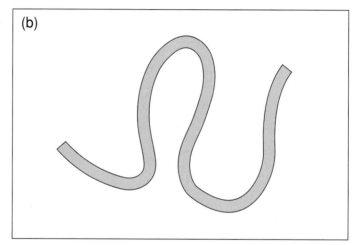

(b)

1. Clay plugs slow bank erosion and bend migration rates. Flow directions change as the radius of curvature of the bend is decreased.
2. Bends are deformed as their downstream limbs are fixed in position by the clay plug (Figure 9.6a). The upstream limb continues to shift downstream compressing the bend and causing a neck cutoff.
3. When the apex of a bend encounters a clay plug, the apex is flattened and in some cases a compound bend develops (Figure 9.6b).
4. If the presence of a clay plug directs the flow into highly erodible adjacent sediments, a very high amplitude bend will develop that will eventually cut off.
5. Numerous clay plugs flanking a channel can inhibit meander development, and a relatively straight channel will be confined to a narrow zone of floodplain (Figure 9.5).

Obviously predictions of meander behavior will be suspect when numerous alluvial inhomogeneities exist. The point of this discussion of the effects of clay plugs is to illustrate how significant river history (prior position) can be to the modern river (Chapter 4). It also shows that care must be used in any attempt to use empirical relations that were developed under ideal conditions of the laboratory flume and computer models. On the Red and Lower Mississippi rivers the majority of bends are influenced by geomorphic and geologic controls and the ideal classic meander is hard to find. Therefore, the task of predicting channel behavior and bank erosion is complicated. In fact, Carson and LaPointe (1983) argue that asymmetry of meanders will be the normal case and experiments (Schumm *et al.*, 1987) reveal that the downstream limb of a meander may be impeded in its downvalley shift by alluvium that is slightly more resistant than the reworked alluvium through which the upstream limb is migrating. This yields a meander like that of Figure 9.6a. Clearly, depending upon bank sediment variability, meanders can be very different from one bend to the next.

Bedrock can also force ground water to the surface when bedrock is close to the surface (Figure 9.3). This produces an interrupted stream (Table 2.1), that is, a local water supply for riparian vegetation. The result is a reach that is narrower and deeper than upstream and downstream reaches where the water table is well below the channel bottom.

Chapter 10

Tributaries

Tributaries are a natural component of a river system (Figure 1.2). If a tributary is small, its impact upon the main channel is small, perhaps causing some minor widening and deepening. However, if the tributary is steep or large it can have effects ranging from significant widening and deepening to metamorphosis, a complete change of channel character.

When a steep sediment-laden tributary joins a main stream, the impact can be substantial (Knighton, 1987). For example, most major rapids along 450 km of the Colorado River in the Grand Canyon (Dolan *et al.*, 1978), are caused by the sediment discharge of steep tributaries that follow faults, normal to the river. Rice and Church (1998) have identified sedimentary links in Canadian gravel-bed rivers that are the result of tributary sediment contribution. These links are a channel reach between two tributaries. The result is a series of links that repeat as a sediment fining sequence. Each link contains sediment that fines downstream with a resulting decrease of gradient. This produces a concave profile segment resulting in a cuspate longitudinal profile.

Rutherfurd (2001) describes the major impact of tributaries on Glenelg River in Victoria, Australia, which has a drainage area of 12 700 km^2. The tributaries supply a very large sand load to the river, which may partly block the river at the tributary junctions (Figure 10.1) and form a back-water lake upstream of the obstruction. The reverse can occur when the main channel aggrades forming lakes in the tributaries (Knighton, 1989).

The introduction of high suspended load from tributaries will change a high width–depth channel to a narrower and deeper channel. As described in Chapter 6, a change of this type occurs along the Smoky Hill River in Kansas (Figure 1.3) which is caused by the introduction of silt and clay from the Saline and Solomon River tributaries. The Smoky Hill River is converted from a high width–depth ratio, low sinuosity stream to a low width–depth ratio highly sinuous river (Figure 6.1) until the Republican River tributary introduces a high sand load at which point the river returns to its original upstream state with a high width–depth ratio (Schumm, 1960).

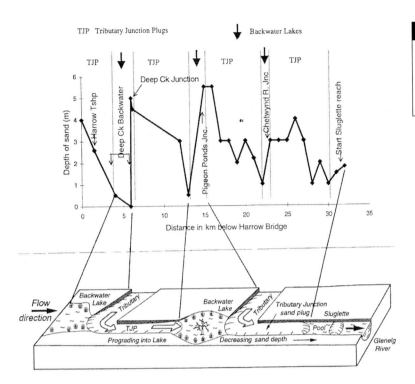

Figure 10.1 Maximum depth of sand in cross-sections surveyed in the Glenelg River Australia, showing the plugs of sand (TJP) formed below the tributary junctions (from Rutherfurd, 2001).

Tributary channels can cause great changes in the main channel, as they modify the flow regime of the main channel. For example, Russell (1954) describes how the classical Meander River of Turkey is affected by the Ak River tributary (Figure 10.2), "The river meanders with some vigor at several places above the junction of the Ak, a tributary from the south that has a profound effect in changing the pattern of the Meander downstream. The Ak is subject to violent floods, and at low water occupies broad, braided channels through gravel and sand bars. The quantity of coarse alluvium brought to the Meander floodplain is sufficient to increase its sandiness for several miles downstream, to account for the development of large sand bars along the Meander channel for a similar distance, and, as a result of the rapidity of bar growth, to introduce a pattern of much larger meander loops along the channel. Sandy islands (towheads) become numerous below the Ak. They create a tendency for the meander to assume a braided course during low stage."

When a tributary causes an increase of gradient downstream from its junction with the main channel, the main channel can adjust by increasing sinuosity. For example, the very sinuous reach of the Mississippi River upstream of the even more sinuous Greenville Bends (Reach 17, Figure 9.4) is associated with valley floor steepening caused by the sediment load introduced by the Arkansas River.

Tributaries can significantly influence longitudinal profiles. For example, the upper Mississippi River valley contains the large Lake Pepin which was formed by glacial outwash deposited at the mouth of the Chippewa River (Figure 1.3) partially blocking the upper

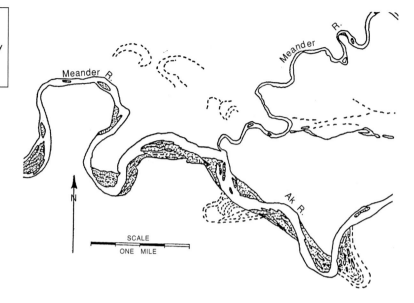

Figure 10.2 Confluence of the Meander and Ak rivers showing the effect of the Ak River tributary on the main Meander River channel (from Russell, 1954).

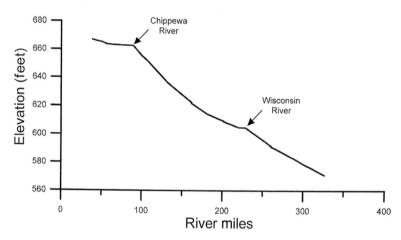

Figure 10.3 Longitudinal profile of a portion of the upper Mississippi River showing effects of Chippewa River and Wisconsin River. Sediment delivered from Chippewa River has partly blocked the Mississippi River to form Lake Pepin (Figure 8.3). River miles, 1. 61 km.

Mississippi River. Downstream to its junction with the Wisconsin River the profile of the upper Mississippi is concave, and it is slightly concave downstream to the bedrock control at Rock Island (Figure 10.3). Both the Chippewa River and Wisconsin River tributaries to the upper Mississippi river significantly influence the longitudinal profile of the river and its valley. In this case, history and tributary influences strongly dominate the river profile.

An extreme example of the effect of a tributary on a large river is the result of the eruption of Mount St. Helens (Bradley, 1983). This eruption on May 18, 1980 delivered three billion cubic yards of rock, ice, and other material into the upper 17 miles (c. 27 km) of the North Fork Toutle valley. The movement of this sediment downstream caused widespread flooding of the Toutle and Cowlitz Rivers, which changed them from meandering to braided. The delivery of large amounts of sediment to the Columbia River blocked its navigation channel and trapped seagoing vessels upstream of the tributary junction.

Another example of the effect of tributaries on a main channel is the shift of the main channel away from the mouth of the tributary to the opposite valley wall. This can occur if the tributary transports sufficient sediment to affect the main channel. For example, many tributaries to the upper Mississippi River have this effect as do tributaries to the Jordan River in Israel where the tributaries build alluvial fans into the valley of the Jordan River (Schattner, 1962). These fans change the gradient of the valley, and according to Schattner (1962, p. 72), downstream of the junction of a large tributary, meanders "tend to be much bigger than those upstream from the junction." In general, tributary effects on the main channel are the result of either a large input of sediment or a change of sediment type.

Chapter 11

Active tectonics

In Chapter 5, the effect of tectonics on sediment yields and river systems was considered. Here, the focus is on local effects of deformation of the valley floor (Figure 1.2). This can include not only active tectonics, but subsidence owing to groundwater and petroleum removal, and local uplift as a result of valley incision (Bell, 1999). These examples can be considered as pseudotectonics, but they can have the same effects as active deformation.

Active tectonics can take several forms. Deformation can be along faults or pairs of faults (horst and graben), which should have the same effect on rivers as a monocline, dome, or basin. In addition, an entire valley may be tilted upstream, downstream, or laterally. The possibilities are great, but, in reality, the primary effect of tectonics will be a local steepening or reduction of gradient or lateral (cross-valley) tilting. In order to adjust to these changes, the river must either degrade, aggrade, or change sinuosity. Small streams in the Mississippi River valley show clearly the impact of active tectonics on their morphology (Boyd and Schumm, 1995; Spitz and Schumm, 1997), and indeed, even large rivers, such as the Mississippi, Indus, and the Nile, change patterns in response to active uplift and faulting (Schumm and Galay, 1994; Schumm and Winkley, 1994). In addition to these primary influences, there will be secondary effects, as rivers respond to changed gradient (aggradation or degradation), and there will be tertiary effects, as decreased or increased sediment loads influence reaches downstream of the deformed reach and as aggradation or degradation in the deformed reach progress upstream.

Therefore, not only will the deformation (fault, fold) affect the channel, but the effects can be propagated upstream and downstream. As noted in Chapter 5, landslides triggered by seismic shock can introduce large quantities of sediment into a river. Bank collapse into the river can greatly increase the sediment load and cause downstream channel adjustments.

Schumm *et al.* (2000) reviewed the ways in which river channels and their sedimentary deposits respond to tectonic deformation (Figure 11.1). The immediate effect will be a local steepening or reduction of gradient. A river may respond by aggrading or degrading,

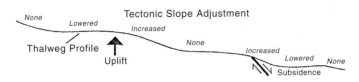

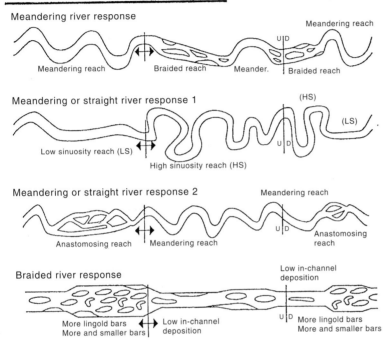

Profile deformation

Common channel pattern adjustments

Figure 11.1 Summary of common channel responses to tectonic deformation (from Holbrook and Schumm, 1999).

and sinuosity, channel shape, and bed material characteristics may change. Conceptually, many combinations are possible, depending on such factors as the rate of deformation, the erodibility of the bedrock into which the channel incises, and the proximity of the channel to a possible pattern threshold.

These responses are closely related to the relation between valley slope and channel pattern with a river changing from straight to progressively more sinuous to transitional meandering-braided (wandering?) to braided (Figure 11.2). If, for example, a straight river is steepened, it should become more sinuous, but at some point, the sinuous channel becomes braided. However, aggradation can cause any type of river to braid.

Valley slope can be affected by faulting, uplift, and subsidence. These changes will affect the channel as noted above, but if the river occupies an alluvial plain, avulsion can be the dominant response (Chapter 5). Cross-valley tilting also may produce effects on channel behavior, such as lateral migration and avulsion (Figure 11.3).

A perfect example of the effect of uplift on a channel 20–55 km above its mouth is provided by Big Colewa Creek in Louisiana where

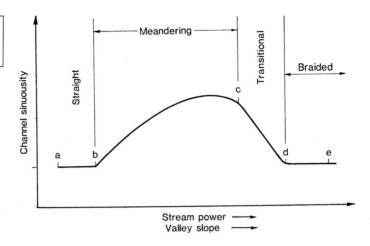

Figure 11.2 Effect of valley slope on sinuosity (after Schumm and Khan, 1972).

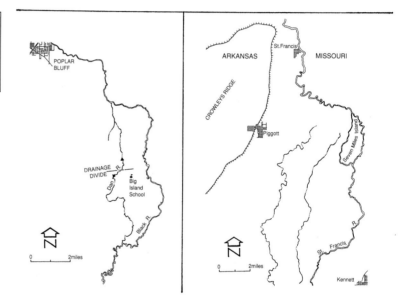

Figure 11.3 Avulsion of Black River and St. Francis River to the east, as a result of tilting of the Mississippi River floodplain (from Boyd and Schumm, 1995).

it crosses the active Monroe uplift (Figures 9.4, 11.4). The channel is actively incising on the axis of uplift, where valley slopes, channel slopes and bank heights increase, and sinuosity increases markedly on the steeper down-valley slope. Other channels that cross the uplift also show a decrease of sinuosity as the river approaches the uplift and then a marked increase on the steeper downstream limb (Schumm *et al.*, 2000).

Larry Lattman (personal communication) provides examples of tectonic effects on medium size rivers based upon his work as a petroleum geologist. The Compton Landing Gas Field underlies the Sacramento River upstream of Sacramento, California. The gas field is related to an active uplift area which has its long axis perpendicular to the river (Figure 11.5). The uplift has warped the valley profile such that the slope has been decreased above the uplift and increased

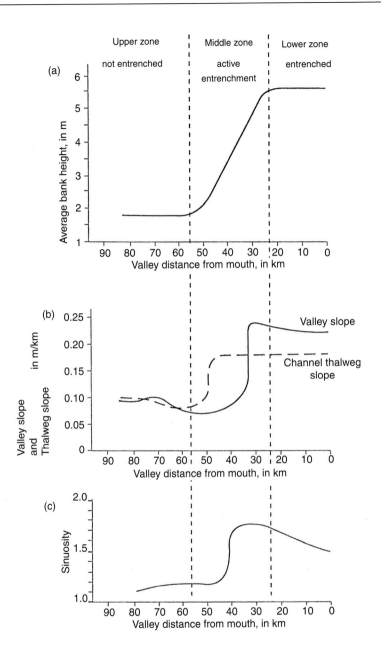

below it. Note the increase of sinuosity downstream of the structural axis. The reduction in slope upstream of the uplift has reduced the velocity of the river and elevated the stage. During floods the river frequently breaches its natural levees and floods the adjacent areas. The ponding of water and breaching of the levee are clearly related to a known structure which is deforming the present valley floor.

The Pecos River crosses a known structure just upstream of the town of Pecos, Texas. The river flows through a broad alluvial valley and it displays remarkable changes in sinuosity (Figure 11.6). The structure causes both a flattening and steepening of the valley floor

Figure 11.5 Map of Sacramento River in the vicinity of the Compton Landing gas field (upstream of Sacramento, CA). Surface expression of structure occurs downstream (south) of gas field location. Dashed lines show extent of overbank flooding.

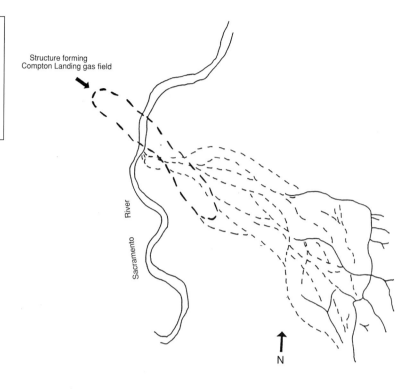

Figure 11.6 Map of Pecos River upstream of Pecos, Texas.

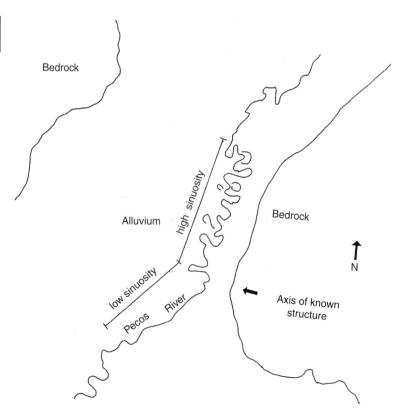

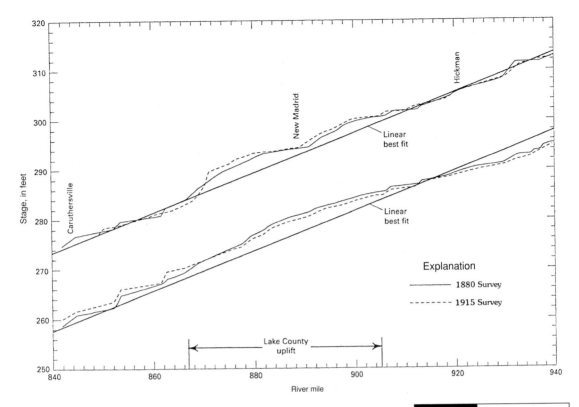

Figure 11.7 Bankfull and medium stages of Mississippi River and location of Lake County uplift (Figure 9.4) as defined by Russ (1982).

similar to what occurs at the Compton Landing Gas Field along the Sacramento River. Where the slope has been decreased, the sinuosity has increased and cutoffs occur frequently. Where slope has been increased, sinuosity has reduced to almost one. This appears to be the reverse of the situations described in Figures 11.1 and 11.2 where gentler valley slopes result in straighter channels, but the straighter reaches of the Pecos River probably are approaching braiding.

Although emphasis has been placed upon variations of planform, it is apparent that other channel characteristics are altered by active tectonics. Gomez and Marron (1991) describe changes in floodplain and channel gradient, sinuosity, floodplain reworking, and channel depth, as a result of deformation, and the examples of the Gulf Coast streams support their findings.

The Lake County Uplift (Figures 9.4, 5.4) has provided a convexity in the profile of the Mississippi River (Figure 11.7) which in turn has, of course, affected not only channel gradient and sinuosity but also channel width (Figure 11.8), depth (Figure 11.9), and width–depth ratio (Figure 11.10). Width and width–depth ratio increase across the downstream part of the uplift, while depth decreases.

Elsewhere, a fault crosses the river at about River Mile 818 (Boyd and Schumm, 1995). The slope of the valley is low upstream, but it steepens markedly at the fault. This steeper reach of the valley floor has a very sinuous channel, and through this steep reach (RM 802–825) width and width–depth ratio decreases and depth increases.

Figure 11.8 Bankfull width of Mississippi River on the Lake County uplift in 1880 and 1903. 1 river mile, 1.61 km. (From Boyd and Schumm, 1995.)

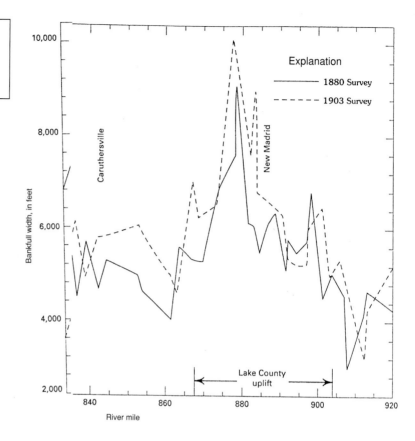

Figure 11.9 Bankfull depth of Mississippi River on the Lake County uplift in 1880 (solid line) and 1903 (dashed line). 1 river mile, 1.61 km. (From Boyd and Schumm, 1995.)

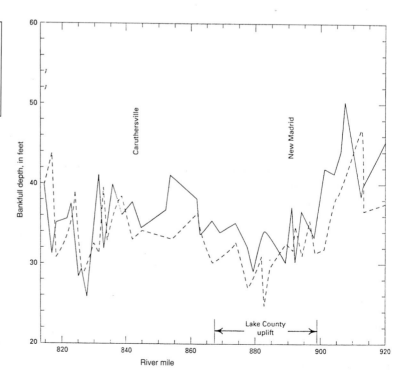

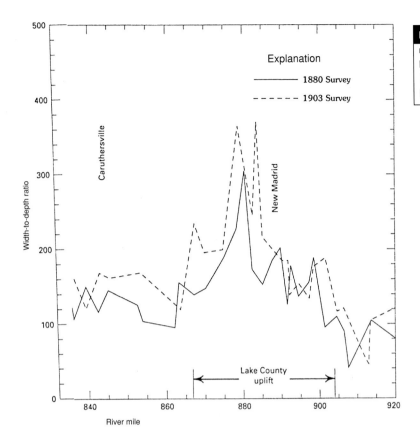

Figure 11.10 Width–depth ratio of Mississippi River on the Lake County uplift in 1880 and 1903. 1 river mile, 1.61 km. (From Boyd and Schumm, 1995.)

These major changes of pattern and dimension clearly reflect changes of valley floor gradient as a result of active tectonics. The Mississippi River and other channels between Cairo, Illinois and Baton Rouge, Louisiana show dramatic effects of valley floor morphology (Schumm and Winkley, 1994; Boyd and Schumm, 1995).

Three streams of the Gulf Coast that cross the Wiggins uplift in southern Mississippi demonstrate that, as might be expected, larger streams more readily adjust to deformation. Therefore, perhaps attention should be directed toward the smaller channels as indicators of active tectonics because the evidence of deformation will be more apparent. For example, Bogue Homo is smaller than Tallahala Creek (Figure 11.11), and the size difference has resulted in different degrees of entrenchment of each creek. Tallahala Creek has undergone significantly more erosion than Bogue Homo, and the depth of its channel through the uplift axis and the quantity of exposed bedrock along the valley course is greater than in the Bogue Homo channel. For example, in a downstream direction along Bogue Homo, the channel becomes anastomosing on the upstream part of the uplift, and incision is primarily on the downstream side of the uplift, whereas the large Tallahala Creek has incised through the axis of uplift and downstream a new floodplain is developing.

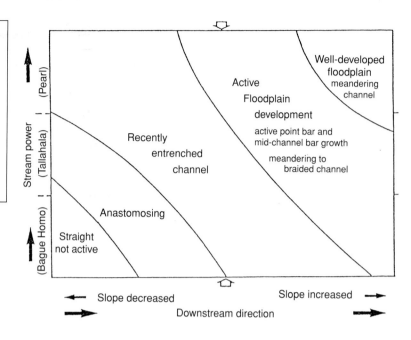

Figure 11.11 Alluvial-river channel characteristics in relation to relative stream power (stream size) and location relative to the axis of active uplift. Note that as stream power increases, the zone of active channel entrenchment moves upstream through the axis of uplift indicating a more complete response to the uplift by larger streams (from Burnett, 1983). Arrows indicate uplift axis.

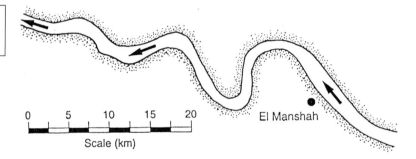

Figure 11.12 Pattern of the River Nile downstream of fault at El Manshah (Figure 4.10).

This evidence indicates that Tallahala Creek and Pearl River adjusted to active uplift to a greater extent than Bogue Homo, presumably due to greater discharge and higher stream power. Although Tallahala Creek may have responded to uplift more than Bogue Homo, both streams presently appear to be in a phase of active adjustment. However, the larger Pearl River appears to be farther along with adjustment to uplift (Figure 11.11). It has incised through the axis of uplift and has a well-developed floodplain downstream. Therefore, the condition of a channel on an uplift will depend on size as well as other factors.

Uplift will also affect flood hydrology. For example, analysis of high-water marks along Tallahala Creek showed that flooding is more likely to occur upstream of an uplift axis, but flooding will be less frequent at and downstream of the axis. This is similar to the Sacramento River situation (Figure 11.5).

Large rivers with great energy, nevertheless, are significantly affected by tectonic alteration of valley slope. For example, the major structural features of the Indus valley seem to dominate the Indus

River pattern (Figure 5.8). Some other reaches of the lower Mississippi River (Figure 9.4) are also affected by deformation. Reaches 3 and 4 are clearly influenced by the Lake County uplift. Reaches 17 through 21 are affected by the Monroe uplift, and faults affect Reaches 5 and 6 and possibly 7, 10, and 12. Clearly, great alluvial rivers can have very significant geologic controls (Schumm and Winkley, 1994).

The River Nile (Figure 4.10), is a remarkably straight river; nevertheless it has short sinuous reaches that are affected by faults that increase the slope of the valley (Schumm and Galay, 1994). For example, the reach downstream of El Manshah is sinuous (Figure 11.12) because a fault steepens the valley at this location.

The discussion has concentrated on straight and sinuous river reaches, but braided rivers, although remaining braided, also respond to changes of valley slope. Germanoski and Schumm (1993) concluded that a braided stream undergoing aggradation upstream of an uplift (Figure 11.1) would increase its intensity of braiding, whereas downstream there would be degradation of the channel, a reduced number of braid bars, and reduced channel width.

Chapter 12

Valley morphology

Examples of valley floor variability (Figure 1.2) and its effects on channels have been given under the discussion of history, tributaries, and tectonics. Convexities and concavities of the valley floor impact a channel, as it incises or deposits, to form a linear longitudinal profile. Many examples show the effect on channel morphology with pattern being the most obvious. The Mississippi, Nile, and Indus rivers provide good examples of this (Figures 5.8, 11.7, 11.12) as discussed in earlier chapters.

However, a different type of channel control exists in semiarid and arid regions. Studies in Wyoming and New Mexico (Schumm and Hadley, 1957) have demonstrated that discontinuous gullies develop in small drainage basins on convexities of the longitudinal profile (Figures 12.1, 12.2). The convexities develop at tributary junctions where the influx of sediment cannot totally be removed and at locations in these dryland valleys where the loss of flow by infiltration causes deposition in long reaches between tributaries (Figure 12.3).

Discontinuous gullies form on the steeper downstream part of the convexity. Depending on sediment delivery to the site, the gullies may be buried or they can extend upstream linking other discontinuous gullies to form a major arroyo (Figure 12.2). Love (1983) shows how different the valleys were in the nineteenth century and at present (Figure 12.4). It has been argued that the concept of discontinuous channels forming on the steeper reaches of a valley would not apply in larger drainage systems, but convexities of the valley floor can be detected in drainage systems as large as the Rio Puerco, New Mexico (Figure 1.3), which is in excess of 140 miles (c. 224 km) long, and it drains an area of 5700 square miles (c. 9120 m^2).

The earliest descriptions of the Rio Puerco valley (c. 1850) indicate that there were discontinuous channels at several locations. For example, Love and Young (1983) state that:

> The entrenchment of the lower Rio Puerco to form the present arroyo apparently took place episodically since at least the 1760s. There was a gully near the present junction of the Rio Puerco and Rio San Jose.

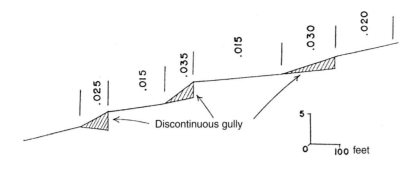

Figure 12.1 Discontinuous gullies in a small well-vegetated valley in Niobrara County, Wyoming. Numbers indicate gradient (from Schumm and Hadley, 1957).

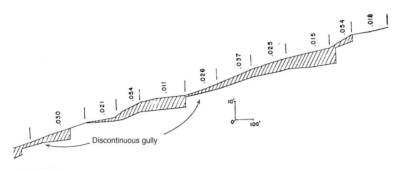

Figure 12.2 Discontinuous gullies near Cuba, New Mexico. Numbers are gradient (from Schumm and Hadley, 1957). Note that two gullies have coalesced.

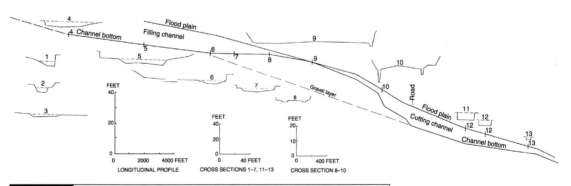

Figure 12.3 Longitudinal profile of portion of Sand Creek, Nebraska showing valley plug of recently deposited sediment between locations 6 and 10. Loss of water by infiltration into the sandy alluvium causes deposition and formation of a broad convexity that blocks the channel and modifies the valley floor (from Schumm, 1960).

In addition, there were discontinuous channels at three other locations. Abert, in 1848 (Bryan, 1928b), described the Rio Puerco with banks of 10 feet (*c.* 3 m) or 12 feet (*c.* 3.7 m) high. Farther upstream, the banks were 30 feet (*c.* 9 m) high. The deepening of the channel upstream is characteristic of discontinuous gullies (Figures 12.1, 12.2). Simpson in 1852 (Bryan, 1928b), described the Rio Puerco at another location as having banks 20 feet (*c.* 6 m) and 30 feet (*c.* 9 m) high.

Figure 12.4 (a) Character of valley floors prior to arroyo formation; sediment can be transported in sheet floods, in a single shallow channel or in multiple channels, but much can be deposited on valley floor.
(b) Character of valley floor after arroyo formation. Locally derived sediments may not reach the incised channel (white arrow) (from Love, 1983).

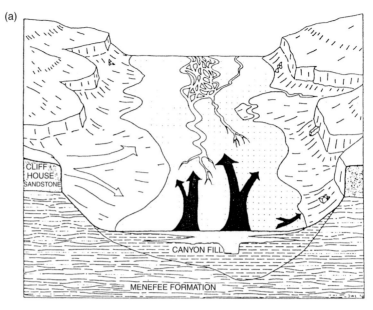

(a)

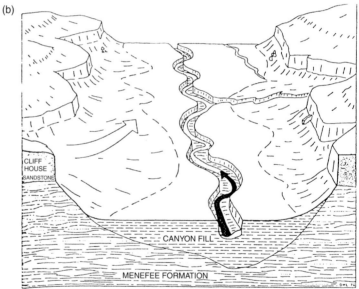

(b)

Downstream near the mouth of Rio Puerco, Garretson, in 1955 (Bryan, 1928b), described the channel as 10–12 feet (c. 3–3.7 m) deep.

Between these discontinuous channels, the Rio Puerco had a shallow channel, and it could be crossed easily. The presence of three or four discontinuous channels indicates that the Rio Puerco in 1848 was unstable, and the coalescence of the discontinuous arroyos would produce a continuous trench upstream from the Rio Grande (Figure 1.3).

In order to determine if the assumption, that discontinuous channels form on convexities of the valley floor in large drainage basins,

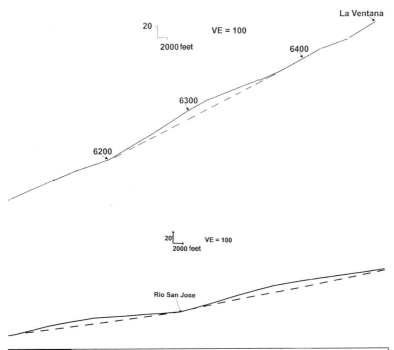

Figure 12.5 Longitudinal profile of Rio Puerco (Figure 1.3) valley showing convexity north of Cabezon, New Mexico. VE, vertical elevation.

Figure 12.6 Longitudinal profile of Rio Puerco valley showing convexities near mouth of large Rio San Jose tributary (Figure 1.3). VE, vertical elevation.

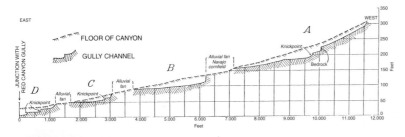

Figure 12.7 Profiles of valley floor and discontinuous gullies in tributary to Red Canyon, Arizona (from Thornthwaite *et al.*, 1942).

is correct, it should be possible to identify steeper reaches of the valley at locations, near the reported discontinuous channels. Indeed, convexities and steep reaches of the longitudinal valley profile are present at these locations and the gradient changes downstream are significant. For example, on a convexity at or near the Simpson location, gradient increases from 0.004 to 0.006 on the downstream part of the convexity. This is a 50-percent increase of gradient (Figure 12.5), sufficient to cause incision and discontinuous gully formation. At the Abert location, there is convexity with gradient increasing from 0.002 to 0.0033, a 65-percent increase.

At the Rio San Jose location (Figure 1.3), there are two convexities (Figure 12.6). On the upstream convexity, gradient increases from

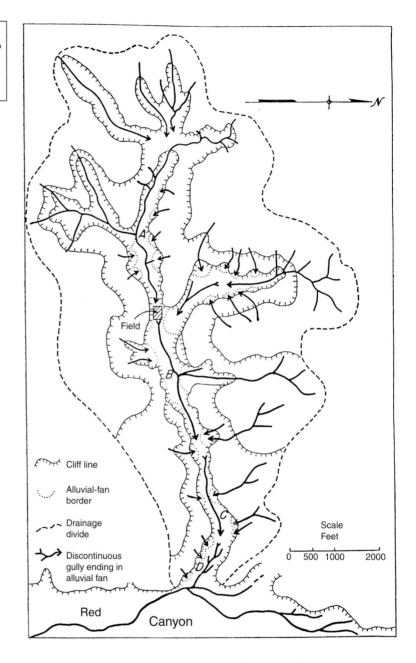

Figure 12.8 Map showing discontinuous gullies in tributary to Red Canyon (Fig. 12.7), a tributary to Polacca Wash in Arizona (from Thornthwaite et al., 1942).

0.002 to 0.003, a 50-percent increase. On the downstream convexity, gradient increases from 0.01 to 0.0025, a major increase. The upstream convexity probably is a backwater effect of the Rio San Jose and the downstream convexity is the result of sediment deposition from Rio San Jose. Near the Garretson location, a convexity changes gradient from 0.0014 to 0.0022, a 57-percent increase of gradient.

In addition to local steepening of the valley floor by sediment deposition, there is a convexity in the cross-section (Figure 12.4). This concentrates flow toward the valley sides, thereby increasing depth of flow and the erosive force of the flow. This would also be the case where a tributary enters a main channel and forces the flow here toward the opposite side of the valley.

A typical series of discontinuous gullies in the Polacca Wash drainage system in Arizona are shown on Figures 12.7 and 12.8. According to Thornthwaite *et al.* (1942), discontinuous gully A has a convex break in its profile (8800 feet or *c*. 2682 m above the junction with Red Canyon). This suggests that the channel is composed of two parts that have recently joined. The presence of a nickpoint 300 feet (*c*. 91 m) downstream from the bedrock waterfall supports this evidence. Above the waterfall the channel is essentially on bedrock throughout its course. Between the mouth of gully A and the main head of gully B (Figure 12.7) is a Navajo cornfield. Gully C contains a nickpoint that is deepening the channel and lowering the gradient as it advances. This is a common occurrence in gullies. Variations in the inherent erodibility of the different horizons of the fill may aid in the formation of nickpoints but are not essential. Gully D, which is continuous with the Red Canyon Gully, has worked headward into the lower part of the alluvial fan formed at the mouth of gully C. Even at the scale of Figure 12.7, it is possible to identify steeper reaches on the canyon floor formed by alluvial fans at 1000, 5000, and 7100 feet (*c*. 305, 1524 and 2164 m).

Figure 12.8 shows another characteristic of these dryland channels. Many tributaries do not join the main channel, rather they disappear on alluvial fans, which eventually are trenched and deliver large amounts of sediment to the main channel.

In the dry climates, where channels carry water only during and after infrequent rains, irregularity of regimen is one of the most marked characteristics of streams. At any part of a channel aggradation in one rain may be succeeded by degradation in the next, and during local showers parts of the stream course may carry heavy run-off while other parts remain dry. There may be many local alternations of erosion and aggradation in dryland valleys and perhaps in rivers in other climatic regions depending upon the quantity of sediment delivered to the channels.

Another effect of valley morphology relates to valley width. Narrow valley sections concentrate flood waters, and therefore, they can have a great impact on the channel, whereas when a valley widens, the flood waters spread widely over the floodplain and the channel may be spared significant modification (Magilligan, 1992).

Part IV

Variable local controls

In addition to fixed local controls there are local controls that are variable (Figure 1.2). That is, they shift position through time, as weather conditions generate local floods and vegetation changes. In addition, accidents such as log jams, fires, and landslides can occur at numerous locations within the drainage network.

Chapter 13

Floods

Floods (Figure 1.2) normally affect almost all of the length of a river but, of course, there are exceptions. For example, the flood may only impact the downstream reaches of a channel or an upstream flood may have its effects dissipated downstream. However, except for extreme events that not only modify the channel, but also the valley, the impacts are ephemeral, being lost as the channel readjusts during floods of lesser magnitude.

Floods produced by local events such as the failure of landslide dams will be discussed in Chapter 15. Here are considered hydrologic events caused by climatic conditions. Wohl (2000b, p. 167) states these conditions as: "A flood may cause dramatic changes along some reaches of a channel and have relatively little effect on other reaches. Similarly, a flood that occurs once every hundred years may create erosional and depositional forms that are completely reworked within 10 years along one channel, but that persists for decades along a neighboring channel."

Stream channels in eastern Australia decrease in size downstream contrary to hydraulic geometry rules (Nanson and Young, 1981a, b). The reason is that the floods go overbank downstream and the channels convey only part of a flood. Also in Australia, Warner (1987a, b) identified periods of higher and periods of lower average rainfall and flooding. These flood-dominated regimes (FDR) and drought-dominated regimes (DDR) tend to persist from 30 to 50 years. The FDR have mean annual discharges from two to four times greater than DDR. During FDR, channel width increases and depth decreases but the greatest effect is along sandy reaches (Pickup, 1986) or where all of the flow is confined to a channel. For example, Mullet Creek (Nanson and Hean, 1985) sustained up to 340 percent widening in confined reaches whereas widening was about 20 percent downstream where the floods were overbank and gradients were gentler. Also, cutoffs were more likely to occur during FDR (Erskine, 1986a, b).

During floods, erosion dominates where velocity and stream power are greatest. For example, during the Big Thompson flood of 1976 in Colorado, channel reaches less than 40 m wide and with gradients of 0.02–0.04 had major scour, whereas reaches greater than

80 m wide and with gradients less than 0.02 had limited scour and widespread deposition (Shroba *et al.*, 1979; Wohl, 2000b). Pickup (1986, p. 112) identified a scour-transfer-fill (STF) sequence where erosion upstream leads to aggradation downstream with, of course, changes of channel morphology from a narrow deep channel to a wide braided channel.

As a result of large floods, the Cimarron River (Figure 1.3) in south-western Kansas was converted from a narrow, deep sinuous channel to a very wide braided channel (Schumm and Lichty, 1963). Subsequent years of smaller floods and more flow caused the channel to revert to a narrower more sinuous channel. These dramatic changes appear to have been confined to southwestern Kansas. The river downstream in more humid eastern Kansas and Oklahoma did not adjust significantly as finer sediments, and vegetation protected the banks.

Obviously, flooding in a tributary could significantly affect the morphology of the main channel downstream of the junction. Kochel (1988) identifies the effects of floods as follows: bank erosion, channel widening, channel erosion, floodplain erosion, channel deposition, floodplain deposition, and bar reorganization. Along any one river, the variability of these processes could provide very different reaches of the channel.

Moss and Kochel (1978) observed great variability of flood effects along Conestoga Creek in Pennsylvania. The flood of Hurricane Agnes caused bank erosion, channel scour, and floodplain modification in the head water stream, where coarse bedload was transported, but in 80 percent of the drainage basin the stream transported primarily suspended sediment, and channel changes were minor. Wolman and Eiler (1958) found that gradient and the ratio of river to valley width controlled scour and deposition in the Connecticut River valley (Figure 13.1).

Floods can significantly modify a channel, but the river itself can determine locations of overbank flooding and floodplain deposition. This occurs along the Alabama River at the apex of active meanders. A shifting meander prevents the full development of natural levees and, therefore, flooding and deposition can be expected more frequently at active bends (Harvey and Schumm, 1994).

Depending upon river action a bank can be relatively stable, eroding or healing. This is the case along the Ohio River (Figure 1.3) where through time, eroding banks heal as the point of attack shifts and vegetation colonizes the bank (Figure 13.2). When there is an absence of large floods, rivers may narrow as vegetation colonizes parts of the channel, and new floodplains are formed. The absence of large floods after 1942 permitted the Cimarron River (Figure 1.3) to narrow from 1200 feet (*c.* 365 m) to less than 100 feet (*c.* 30.5 m). There can be no doubt that large floods and especially floods of long duration can cause bank erosion and channel shift. Of course, flood effects are greatest in steep gradient channels. Other examples of flood effects are provided by Stevens *et al.* (1975b), Graf (1983), Baker *et al.* (1988), Kochel (1988), and Everitt (1993).

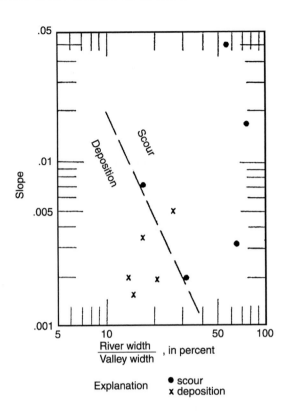

Figure 13.1 Conditions of deposition and scour related to slope and ratio of river width to valley width (from Wolman and Eiler, 1958).

The Gila River in Arizona provides a good example of channel adjustment to floods. Burkham (1972) studied the changes of the Gila River in the Safford Valley of southeastern Arizona for the period 1846–1970. The changes can be grouped into three periods: 1846–1904, 1905–17, and 1918–70. From 1846 to 1904, the stream channel was narrow and meandered through a flood plain covered with willow, cottonwood, and mesquite. Only moderate changes occurred in the width and sinuosity of the stream channel in this period; the average width of the stream channel was less than 150 feet (*c.* 46 m) in 1875.

The average width of the channel of the Gila River increased during 1905–17 to about 2000 feet (*c.* 610 m) mainly as a result of large winter floods that carried relatively small sediment loads. The meander pattern of the stream and the vegetation in the flood plain were destroyed completely by the floods. However, with smaller floods, the channel of the Gila River narrowed during the period 1918–70, and the average width was less than 200 feet (*c.* 61 m) in 1964. The channel developed a meander pattern and the flood plain became densely covered with vegetation. Minor widening of the stream channel occurred in 1965 and in 1967, and the average width of the channel was about 400 feet (*c.* 122 m) in 1968. The reconstruction of the flood plain in 1918–70 was accomplished almost entirely by the vertical accretion of sediment. The most important factors influencing the deposition of sediment in the Gila River during 1918–70 were the wide channel

(a)

(b)

(c)

Figure 13.2 Bank characteristics of the Ohio River (Figure 1.3). (a) Eroding bank; (b) healing bank; and (c) stable bank.

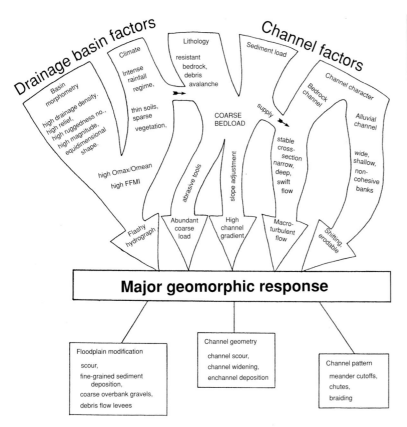

Figure 13.3 Factors controlling channel and floodplain response for large floods (from Kochel, 1988).

and the small floods that carried relatively large sediment loads. The large sediment loads resulted mainly from rapid erosion of the alluvial deposits in the drainage basins tributary to the Gila River.

Huckleberry (1994) stresses duration of flooding as a cause of Gila River widening during a 1993 flood. He contrasts the effects of two large floods, which occurred in 1983 and 1993 on the Gila River. The 1983 flood had a peak of about 100 000 cfs (c. 2830 m^3/s). In 1993 the peak discharge was 74,900 cfs (c. 2120 m^3/s). The 1983 flood lasted for about two days, whereas in 1993 "there was prolonged high discharge with several peaks associated with individual fronts; average daily discharges exceeded ~280 m^3/s (100 000 ft^3/s) over a period of two weeks" (Huckleberry, 1994). Accordingly, little change of the channel occurred in 1983, but great change resulted in 1993. Channel width–depth ratios increased as much as eight times, and the channel became wide and braided.

It is important to recognize that not only is the distribution of precipitation within a watershed important for where flood impacts occur, but the nature of the floods are significant. For example, where the ratio of peak flood discharge to average annual discharge is small the river can, according to Stevens *et al.* (1975b) be in regime, whereas when the ratio is large, the channel will exhibit nonequilibrium form.

This ratio for the lower Mississippi is about 1.7 whereas for the Gila River it is about 10.

Kochel (1988) summarizes the various factors that determine the impact of floods on rivers (Figure 13.3). Drainage basin morphology and climate combine to yield a flashy hydrograph. Resistant rock yields coarse sediment load and this produces a steep gradient and an alluvial channel that is highly adjustable. The result is large, flashy floods that move coarse bedload in an adjustable channel with significant modification of the floodplain, channel dimension, and pattern. Obviously, a river can have significant morphologic changes downstream, depending on the location of the source of flood waters, the magnitude of the flood, as well as its duration.

Chapter 14

Vegetation

It appears intuitive that any vegetation (Figure 1.2) on the banks of a river will protect that bank from erosion. Numerous field studies support this contention. For example, following major floods in British Columbia it was determined that of 748 bends, those without riparian vegetation were five times as likely to have eroded, and major bank erosion was 30 times more prevalent on non-vegetated bends (Beeson and Doyle, 1995). Smith (1976) found in cold regions that heavily vegetated banks were 20 000 times more resistant to erosion than non-vegetated banks. Ikeda and Izumi (1990) state that channels with dense vegetation will on average be 30 percent narrower than channels with little or no vegetation. However, if banks are high and they exceed the depth of roots, undercutting will cause bank failure. In addition, the weight of trees will accelerate failure especially if wind is a factor.

Hey and Thorne (1986) developed a series of equations relating channel width to discharge for different bank vegetation densities (Figure 14.1). Bank vegetation ranged from 0 percent trees and shrubs (Type 1), 1–5 percent (Type 2), 5–50 percent (Type 3), and greater than 50 percent trees and shrubs (Type 4):

$$
\begin{aligned}
&\text{Type} \quad 1 \; w = 4.33 \; Q^{0.5} \\
&\text{Type} \quad 2 \; w = 3.33 \; Q^{0.5} \\
&\text{Type} \quad 3 \; w = 2.73 \; Q^{0.5} \\
&\text{Type} \quad 4 \; w = 2.34 \; Q^{0.5}
\end{aligned}
\tag{14.1}
$$

Clearly, channel width was greater when trees and shrubs were lacking. Similar results were documented for small streams in New South Wales (Huang and Nanson, 1997). Along the banks of the Genesee River in New York, bends that encountered farmland migrated 30 percent faster than the bends that encountered forest land (Beck *et al.*, 1984) and cultivated and fallow fields yield channels with larger cross-sections than forested areas (Figure 14.2).

An experimental study of a braided channel (Gram and Paola, 2001) demonstrated that vegetation plays an important role in stabilizing the banks, constraining channel migration, and causing

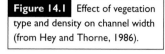
Figure 14.1 Effect of vegetation type and density on channel width (from Hey and Thorne, 1986).

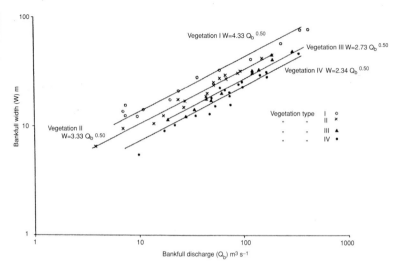

Figure 14.2 Channel cross-section changes with drainage area for three types of vegetation in Nigeria (from Odemerho, 1984).

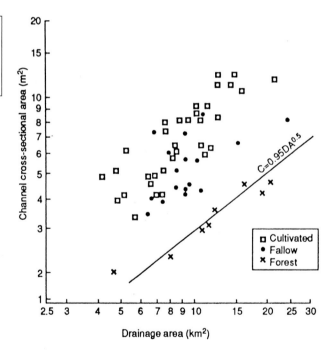

channels to deepen and narrow. Millar (2000) describes the changes taking place as a result of logging to the banks of Slesse Creek a fourth-order stream in British Columbia. Prior to logging, the channel was slightly sinuous with a width of 30 m. After logging, the channel widened to about 150 m, and it braided. Conversely, encroachment of woody vegetation into the channel of the Platte River (Australia) has significantly narrowed the channel (Figure 7.6). In fact, the vegetational colonization of a wide braided channel in central Australia caused preferential trapping of sediment, which eventually formed

(a)

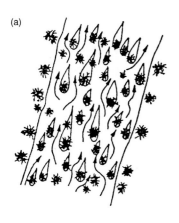

Chaotic growth of teatrees and associated ridges in channel. Ridges approximate streamlined (lemniscate) forms.

Figure 14.3 Diagrams illustrating three stages of ridge formation as a result of colonization of an Australian channel by trees (from Tooth and Nansen, 2000).

(b)

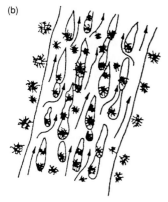

Teatrees and ridges start to become ordered into stream-parallel groves. Ridges begin to deviate from the lemniscate form.

(c)

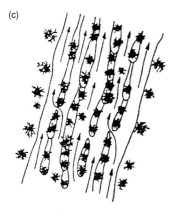

Ridges coalesce at their upstream and downstream ends, and separate anabranches relatively free of teatrees or small ridges. Ridges deviate considerably from the lemniscate form.

linear ridges (Figure 14.3) that produced an unusual anastomosing channel (Tooth and Nanson, 2000).

However, in spite of the preponderance of data supporting the conclusion that vegetation controls bank stability, anomalous results have been obtained by Trimble (1997) and Hession et al. (2003). These investigations found that for small watersheds in Wisconsin, Pennsylvania, Maryland, and Delaware, channels with forested banks are wider than grassed banks. For example, Hession et al. (2003) found that width varied with drainage areas for forested banks: $w = 4.15A^{0.30}$;

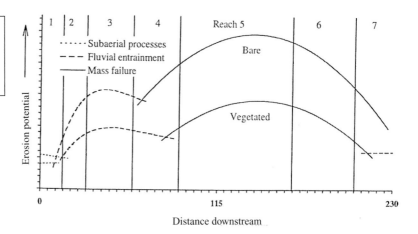

Figure 14.4 Changing type of erosion and erosion potential in a downstream direction for Latrobe River, Australia (from Abernethy and Rutherfurd, 1998).

whereas for grassed banks: $w = 1.97A^{0.46}$. An explanation is that grassed banks are able to trap sediment and cause channel narrowing. Another explanation is that large woody debris in the forested reaches causes deflection of flow toward banks and locally destabilizes the banks.

A very important factor is the drainage area supplying water and sediment for a reach. According to a study of the changing effect of riparian vegetation along the Latrobe River in southern Victoria, Australia (Abernethy and Rutherfurd, 1998), vegetation has different impacts on stream morphology depending upon location in the drainage network (Figure 14.4). In the uppermost reach of the rivers, channel size and morphology are influenced by wind throw of trees, damming by woody debris, frost heave, and surface erosion of exposed banks. In the middle reaches 2–4 of the river (Figure 14.4), fluvial erosion is dominant as stream power increases and the size of bank sediments decrease. Here vegetative cover is important. In the downstream reaches (5, 6), bank height attains a critical condition and mass failure dominates. In all reaches, bare banks erode and fail more than vegetated banks, although the processes of erosion are very different (Figure 14.4).

Clearly, variability of bank vegetation causes variability of channel morphology and removal of vegetation naturally or by human activity can produce significant channel change. Depending upon a point of view and the objectives of channel work, the morphologic changes may be desirable. For example, Henry Shreve, while clearing the log jams (see Chapter 15) from the Red River (Figure 1.3) in Louisiana, concluded that the best way to maintain an open navigable channel was to clear-cut riparian forests before the trees became snags. Shreve instructed his men to fell every tree within 300 yards (c. 274 m) of the bank. During the height of the program, from November 1842 to July 1845, government snag crews removed about 80 000 riverfront trees (Shallat, 1994). This type of activity suggests an explanation for the marked increase of width of the lower Mississippi River between 1821 and 1880 (Figure 14.5). During this period, cutting of trees to

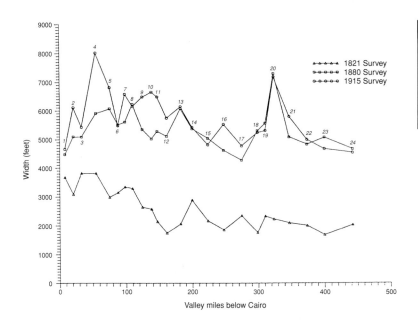

Figure 14.5 Average bankfull width of Mississippi River for 24 reaches (Figure 9.4) in 1821, 1880, and 1915. Note major increase of width between 1821 and 1880 (from Schumm and Winkley, 1994).

fuel the numerous steamboats on the river was a factor in causing bank erosion and dramatic channel widening (Winkley, 1977).

Depending on other factors, vegetation can be of great or of little significance regarding bank stability and channel type and variation of vegetation can locally cause significant changes of channel morphology downstream.

Chapter 15

Accidents

Accidents (Figure 1.2), as discussed here, are unanticipated events. Most floods are not included (see Chapter 13) because they can be anticipated. For example, 100-year floods do eventually occur. Accidents are earthquakes, log jams, ice jams, avalanches, and volcanic eruptions, which are not normally anticipated and yet they produce dramatic channel changes and catastrophic floods.

Log jams and snags

Trees that fall into a channel can have a marked effect. In small channels, single trees can obstruct flow, thereby causing a back-water effect and sediment deposition. When the trees rot or are otherwise removed, incision into the sedimentary deposit will produce discontinuous terraces (Figure 15.1). Similar results occur as a result of beaver dam construction (Bigler, 2001). In a sinuous channel, the back-water effect of log jams can cause cutoffs. In mountain streams evidence of former jams is a major increase of channel width and the presence of mid-channel bars (Keller and Swanson, 1979).

Even larger rivers in forested regions can be blocked by numerous fallen trees (snags) causing an anastomosing pattern. For example, in 1872 the Willamette River in Oregon frequently had four or five channels. Unlike today's trees, the size of snags in the Willamette River ranged from 30 to 60 m long and up to 2 m in diameter. Clearing of the snags produced a single channel (Sedell and Froggatt, 1984).

Large log jams of up to 8 km long were common on many US rivers in the nineteenth century. Table 15.1 illustrates the magnitude of this problem. Even when the rivers were not blocked, the snags were a serious hazard to navigation. In the Ohio and Mississippi Rivers, snags that were anchored in the sediments of the river bottom acted as spears that could pierce the hull of the river boats. Henry Shreve designed and built a snag boat that successfully cleared the snags in

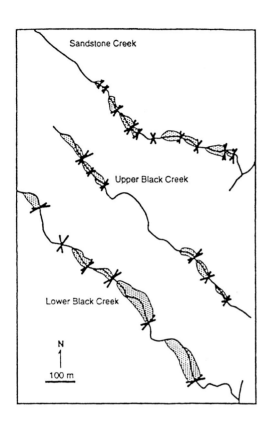

Figure 15.1 Maps of
southeast-flowing Black and
Sandstone creeks showing the
distribution of log jams (crosses)
and alluvial terraces (stippled)
(from Montgomery et al., 1996).

the Ohio and Mississippi Rivers (McCall, 1984). However, his greatest
challenge was to clear the great Red River Raft, a 160- to 200-mile
(c. 49–61 m) long log jam that prevented navigation on the Red River
(Figure 1.3) in Louisiana.

Humphreys and Abbot (1876) describe this feature as: "This raft is
composed of an immense accumulation of drift logs – some floating,
and others so water-logged as to sink, and thus still more effectually
block up the channel. From the rotting of the logs at the lower end,
and the fresh accessions at the upper end, the Raft gradually moves
upstream" (Figure 15.2).

Shreve with his snag boats removed the Raft during the period
1833 to 1839. However, log jams continued to occur until 1873. The
effect of the Raft was to produce multiple channels and numerous
lakes and the removal of the Raft caused major channel adjustments.
A single channel carried the flow after Raft removal, which caused
a 2.7 increase of channel width from 509 feet (c. 155 m) in 1886 to
1251 feet (c. 381 m) in 1938. Channel depth increased from 37 feet
(c. 11 m) in 1886 to 40 feet (c. 12 m) in 1938 (Harvey et al., 1988).

Log jams on the main channel can have major impacts on tributaries by raising local base-level and forming tributary lakes (Figure
15.3). It appears that removal of large woody debris from many rivers
has greatly modified their channels.

Table 15.1 Summary of snags pulled from rivers in the United States for navigation improvement from 1867 to 1912

Rivers by region	Period of snagging	Kilometers snagged	Snags removed
Southeast region			
Pamunkey R., Virginia	1880–1912	50	3677
North Landing R, North Carolina and Virginia	1879–1897	28	9012
Pamlico and Tar R., North Carolina	1879–1912	81	29 260
Contentnia Cr., North Carolina	1881–1912	116	10 372
Black R., North Carolina	1887–1912	116	11 685
Edisto R., South Carolina	1882–1906	124	26 512
Savannah R. to Augusta, Georgia	1881–1912	409	37 812
Oconee R., Georgia	1877–1912	163	44 840
Noxubee R., Alabama and Mississippi	1890–1901	114	143 700
Pearl R., Mississippi	1879–1912	744	294 300
Tombigbee R., Mississippi	1892–1912	794	286 220
Guyandot R., West Virginia	1890–1899	134	8060
Cumberland R., above Nashville, Tennessee	1879–1908	591	38 828
Choctawhatchee R., Florida and Alabama	1874–1912	350	177 599
Oklawaha R., Florida	1891–1911	102	9089
Caloosahatchee R., Florida	1886–1911	36	7894
Central region			
Grand R., Michigan	1905–1911	67	2019
Minnesota R., Minnesota	1867–1912	396	13 740
Red R., North Dakota and Minnesota	1877–1912	528	3600
Red Lake R., North Dakota and Minnesota	1877–1912	248	1500
Wabash R., Illinois and Indiana	1871–1906	79	7700
Missouri R.	1879–1901	2888	25 030
Arkansas R.	1879–1912	1980	139 214
White R., Arkansas	1880–1912	495	22 500
Cache R., Arkansas	1888–1912	162	26 030
St. Francis and L'Anguille R., Arkansas	1902–1912	36	6700
Southwest region			
Guadalupe R., Texas	1907–1912	86	70 583
West Coast region			
Sacramento R., California	1886–1920	380	33 545
Chehalis R., Washington	1884–1910	25	4838
Willamette R., above Albany, Oregon	1870–1880	91	5362

(From Harmon *et al.*, 1987.)

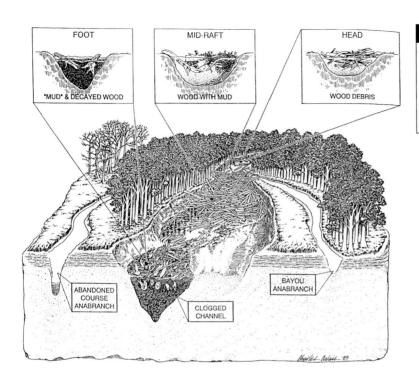

Figure 15.2 An artist's conception of the Great Raft on the Red River in Louisiana showing the foot, midraft, and head of the raft. Note the overflow anabranch and distributary channels (from Albertson and Patrick, 1996).

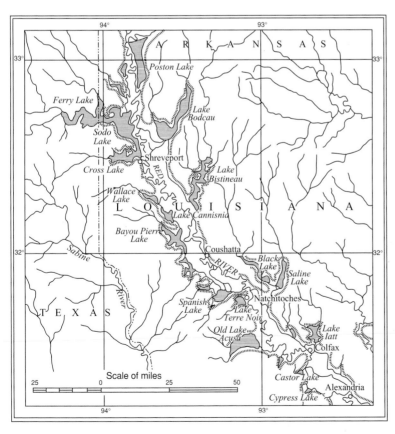

Figure 15.3 The lakes of Red River Valley in Louisiana at their fullest recorded development (from Veatch, 1906). Most are now drained.

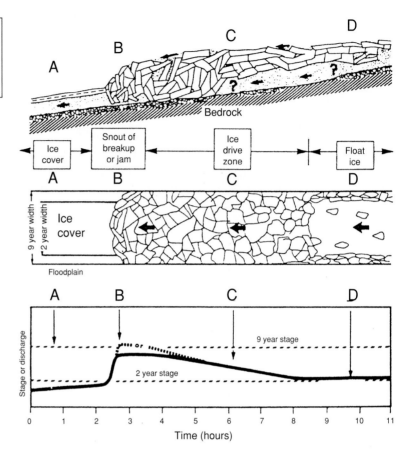

Figure 15.4 Idealized diagram showing the stages of ice break-up and drive over distance and time for a typical large northern river (from Smith, 1979).

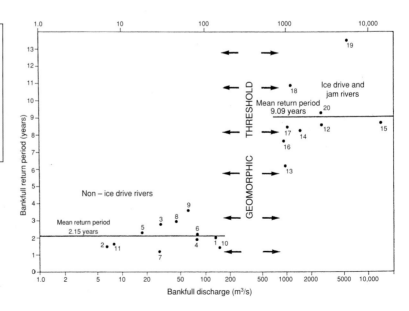

Figure 15.5 Using data on summer floods, bankfull discharges and return periods were calculated for small and large rivers in Alberta. Note that a two-year flood fills a non-ice drive channel whereas a nine-year flood is required to fill channels subjected to ice drives (from Smith, 1980).

Ice drives

Similar to log jams are ice jams, which temporarily block channels and cause overbank flooding and deposition. More significant for channel morphology are ice drives. An ice drive involves the flushing away of surface ice in a channel (Figure 15.4). Some drives are spectacular. A period of warm weather or early spring rain with a rise of discharge can cause a sudden breakup and an ice drive. The impact is to widen the channel (Figure 15.5). The widened channel cross-section at bankfull on average is 2.6 times larger than the non-ice flow channels (Smith, 1979), and the return period of bankfull floods in the ice-drive channel is 4.5 times longer from 2.2 to 9.1 years (Smith, 1980).

Natural dams

Humans commenced dam construction as early as 3000 BC in Jordan and 2600 BC in Egypt (Schnitter, 1994), but natural dams caused by blockage of valleys by landslides, glaciers, and lava dams have created hazardous conditions downstream because these dams fail (Cenderelli, 2000).

Earthquakes, which cause massive landslides, produce about 40 percent of natural dams. For example, following an 8.6 magnitude earthquake in Assam, the River Subansiri, a tributary to the Brahmaputra, was dammed by a landslide which was 40 m high. The dam failed after four days, inundating about 1000 km^2 downstream (Bapat, 1988). A similar situation occurred in New Zealand on the Buller River (Adams, 1981a, b). If a river is not dammed then rapids may persist at the landslide site (Hammack and Wohl, 1996; Webb et al., 1999), but it is the failure of the natural dams that generate major floods.

Most natural landslide dams result from excessive precipitation, both rainfall and snow melt (Costa and Schuster, 1988) and most fail shortly after formation. For example, Schuster and Costa (1986) using a sample of 63 cases determined that 22 percent failed in less than one day and 50 percent failed within 10 days.

The natural dams trap sediment upstream, and after dam failure this deposit may remain, although dissected as a delta, thereby changing channel dimensions and gradient. Failure of the dam flushes the stored sediment and the landslide material downstream causing aggradation and braiding which produces a wide shallow reach of channel (Schuster and Costa, 1986, p. 17).

A major landslide in Jackson Hole, Wyoming dammed the Gros Ventre River on June 23, 1925. The dam was 225 feet (c. 69 m) high and nearly one-half mile (800 m) wide. On May 18, 1927 the dam was overtopped causing a sudden disastrous flood which caused considerable destruction and some loss of life downstream. The meandering gravel-bed Gros Ventre River was converted to a wide braided river (Mulhern, 1975).

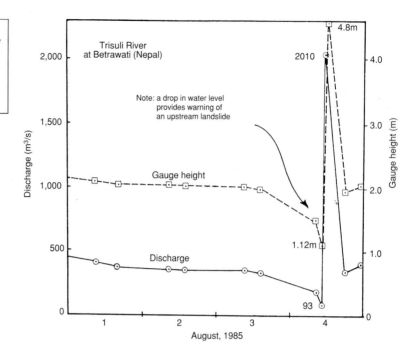

Figure 15.6 Changing gage height and discharge, as a result of the formation and failure of landslide dams, Trisuli River at Betrawati, Nepal (from Galay, 1987).

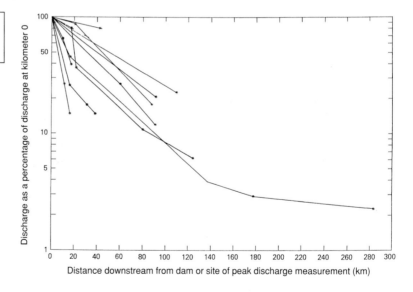

Figure 15.7 Attenuation rates of floods resulting from dam failures (from Costa, 1985).

In December 1840, a massive earthquake-generated landslide dammed the Indus River in the Himalayas of northern Pakistan. The dam was probably about 220 m high. In May or June 1841 the dam was overtopped, and the lake that had formed behind the dam and which extended for 60 miles upvalley drained catastrophically causing tremendous damage downstream. An army camped close to the river at Attock 420 km downstream was overwhelmed by a flood-wave estimated to be 25 m high. About 500 soldiers were killed (Schuster and Costa, 1986; Code and Sirhindi, 1986).

A landslide upon blocking a river causes a rapid decrease of discharge and, of course, gage height (Figure 15.6) which should provide a warning downstream before the dam fails and releases large quantities of water. Fortunately, flood surges are attenuated rapidly downstream (Figure 15.7). Therefore, the effects will be confined to about 100 km downstream.

Accidents can completely change the character of a river, but the downstream effect can be limited to a short reach or to a very long one, depending upon the magnitude and type of accident.

Part V

Downstream controls

There is really only one downstream variable, and it is gradient change. However, the change occurs in four ways. A rise or a fall of base-level increases or decreases the gradient. The same impact on a channel occurs as the channel lengthens or shortens either naturally or by human influences.

Chapter 16

Base-level

It seems very logical that base-level lowering (Figure 1.2) will rejuvenate a drainage network, deliver large amounts of sediment downstream, and cause major channel change. Support for these conclusions comes from experimental studies (Holland and Pickup, 1976; Schumm *et al.*, 1987) and from observations of the headward incision of arroyos, gullies and channelized streams (Schumm *et al.*, 1984). Fisk (1944) concluded that Pleistocene sea-level lowering caused excavation of alluvium and bedrock incision by the Mississippi River for a very long distance upstream. On a smaller scale even a single meander cutoff can cause local steepening and upstream channel degradation (Winkley, 1977). Lane (1955) an eminent river engineer concluded that, when a river is affected by a base-level rise or fall (Figure 16.1), the river will degrade or aggrade to restore its equilibrium profile. After all, the channel must continue to carry its load of sediment with a given water discharge, and this requires a given gradient that must be restored either by deposition (Figure 16.1a) or by erosion (Figure 16.1b).

In contrast to Lane, Leopold and Bull (1979) concluded that base-level changes affect the vertical position of a river only locally and to a minor extent. They argued that not only is stream gradient change important, but that channel pattern, roughness and shape (Rubey, 1952) can also adjust, in order to absorb the effect of base-level change.

Saucier (1981) concluded from the extent and shape of the prism of Holocene backswamp deposits in the Mississippi River valley, that the direct influence of the late-Wisconsin sea-level fall (120 m) extended no farther than 370 river km upvalley (Autin *et al.*, 1991, p. 552). This is only a fraction of the distance to the head of the Mississippi alluvial valley at Cairo, Illinois, 1536 river km upstream. Nevertheless, an effect extending 370 km upstream is not local and minor. However, Pleistocene sea-level lowering caused only 23 m of incision by the Trinity and Sabine rivers on the Texas continental shelf (Thomas and Anderson, 1989, p. 566). The subsequent sea-level rise caused deposition for a distance of only 150 km upstream (Thomas and Anderson, 1989). Blum's (1992) studies of the Colorado River of Texas led him to a similar conclusion. Further investigation shows that there was

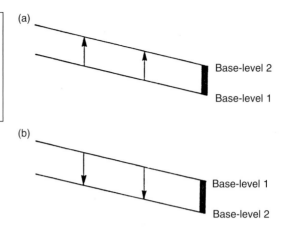

Figure 16.1 When base-level changes, the stream profile will be established at a higher or lower elevation (a) base-level rises from position 1 to 2, (b) baselevel falls from position 1 to 2 (after Lane, 1955).

incision to a depth of 40 m (Thomas and Anderson, 1989), but this was far less than the 120 m of sea-level lowering (Fairbanks, 1989).

The question regarding the impact of base-level change on the rivers does not have a ready answer. If the base-level effect is minimal in some cases, but not in others, then other variables must play an important role. At least 10 variables appear significant, and they can be grouped into three classes: (1) base-level controls, including direction, magnitude, rate, and duration of the change; (2) geologic controls, including lithology, structure, and nature of valley alluvium; and (3) geomorphic controls, including inclination of exposed surfaces, valley morphology, and river morphology and adjustability.

The direction of base-level change obviously determines whether a river will aggrade or degrade, but the magnitude of the change appears to be most important. If base-level lowering is small, a channel can adjust to a change of slope by changing its pattern, by increasing bed roughness, or by changing shape, and in this manner it can accommodate a base-level change. If the change is large, river incision is the result.

The rate of change is also important, and some experimental work has demonstrated that when the lowering of base-level is fast, a stream incises vertically with little lateral migration, whereas when the change is slow, considerable lateral migration takes place, permitting the channel to adjust its slope (Yoxall, 1969; Wood et al., 1993). Also with rapid incision, all discharge events, even the largest flows, will be concentrated in a narrow deep channel. This increases the energy of the flow, and the effect of the base-level lowering is greatly enhanced. In alluvial valleys in the southeastern United States much of the flow was spread over the valley floor prior to channelization, and this flow was, in effect, non-erosive. Following initial rapid incision, floodwaters were concentrated in the enlarged channel, and incision moved rapidly upstream (Schumm et al., 1984). Therefore, rapid base-level change appears to have a more significant upstream impact, than does a slow change.

The cohesiveness of the sediment forming the bed and banks seems to be a very important factor in determining how far upstream the effect of a base-level lowering can propagate. Experimental studies in low-cohesion sediments have shown that nickpoints will not migrate indefinitely upstream (Brush and Wolman, 1960) because: (1) as the nickpoint reach lengthens, its slope is reduced until it is nearly equal to the slope of the stream; the nickpoint cannot be identified when the slope of the nickpoint reach is approximately 20 percent of the average slope of the river; (2) as slope reduction takes place, stream competence declines to the point that bedload movement ceases; and (3) considerable bank erosion takes place in cohesionless sediment so that widening of the channel occurs, stream competence declines, and the channel braids.

Lane's (1955) argument that a stream will restore its gradient at a higher or lower level following a base-level change (Figure 16.1) is conceptually correct because if the river is initially at grade, then in order to move its sediment load and water discharge through the channel, the original gradient of the stream must be re-established. However, sediment loads may be greater after channel incision and less after aggradation. A fallacy in Lane's argument is the assumption that valley slope, and channel gradient are identical. This is not the case and the two-dimensional perspective provided by Figure 16.1 leads to erroneous conclusions. For example, a sinuous river has a gradient less than that of the valley floor.

Sinuosity (P) is the ratio between channel length (Lc) and valley length (Lv), and it is also the ratio between valley slope (Sv) and channel gradient (Sc):

$$P = \frac{Lc}{Lv} = \frac{Sv}{Sc} \qquad (16.1)$$

Therefore, a straight channel has a sinuosity of 1.0, and the gradient of the channel and the slope of the valley floor are the same. It has been demonstrated that rivers can respond to major changes of water and sediment load primarily by pattern changes (Schumm, 1968), and that much of the pattern variability of large alluvial rivers such as the Mississippi, Indus, and Nile reflect the variability of the valley slope.

Figures 16.2 and 16.3 are an attempt to illustrate this concept by simple geometric figures. Figure 16.2 shows the impact of the lowering of base-level in a valley with a stream of sinuosity (P) of 1.5. The line A–C represents channel slope and the line A–B represents the slope of the valley floor. Points B and C are at the valley mouth (B) and the river mouth (C) and points F and G are at the same location in the valley. The channel distance is one-third longer than the valley distance, and the difference in channel and valley slope is reflected in the sinuosity of the stream. The length of the channel is 1.5 times the length of the valley and, therefore, the stream gradient is one-third less than the valley slope (Equation 16.1). If a vertical fall of

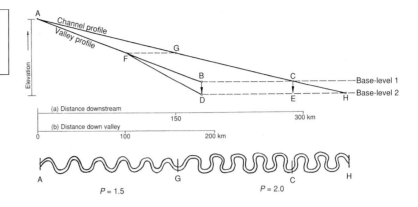

Figure 16.2 Effect of a base-level fall (B to D) on channel length and pattern. See text for discussion (from Schumm, 1993).

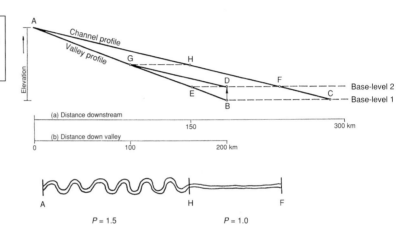

Figure 16.3 Effect of base-level rise (B to D) on channel length and pattern. See text for discussion (from Schumm, 1993).

base-level from B to D and C to E is assumed, channel incision and lateral erosion will steepen the valley floor (F–D). If the channel is not confined laterally, it can adjust to the increased valley slope (F–D) by increasing sinuosity to 2.0, as the channel profile is extended to H. In this case, incision ceases at point F in the valley and at point G in the channel, because the increase of sinuosity from 1.5 to 2.0 from G to H maintains a constant channel gradient over the reach of increased valley slope (F–D). The one-third increase of channel length by increased sinuosity between G and H compensates for a one-third steepening of the valley floor from F to D.

According to Lane's assumptions, the effect of this base-level fall will be propagated upstream to point A where an amount of erosion equal to B–D will occur. However, because the stream can adjust, the steepening of the valley floor will not result in the change of stream gradient. Rather the channel lengthens to accommodate the change, and the effect of base-level lowering is propagated only a relatively short distance upstream. The distance will undoubtedly depend on local conditions and the original slope of the valley floor, but this exercise supports Saucier's (1994) contention that Pleistocene sea-level change in the lower Mississippi valley was effective only upstream as far as Baton Rouge. The probability that a large river can adjust in this fashion is made more likely by the fact that the base-level changes in nature will take place relatively slowly and not abruptly. The river, therefore, has more time to adjust by changing sinuosity.

Figure 16.3 shows an abrupt rise in base-level from B to D for a valley profile A–B and a channel profile A–C. Under Lane's assumptions, deposition would backfill the valley causing deposition equal to B–D at point A. This situation, identical to deposition behind a dam, will produce deposition, but when the valley aggrades to a slope (G–D) equal to the original channel gradient (A–C), which is necessary to transport water and the sediment load to the new shoreline at point F, deposition should cease. On the reduced valley slope (G–D) sinuosity will decrease from 1.5 (A–H) to 1.0 (H–F). In this way, the gradient of the channel is maintained, and deposition ceases at point G in the valley and point H in the channel. In this case, the effect of an abrupt rise in base-level was not propagated upstream to point A because the effect was absorbed by a decrease of sinuosity from H to F.

Figures 16.2 and 16.3 show only two examples of base-level change, where the sinuosity of the river before the base-level change was 1.5. However, sinuosity could range from 1.0 to about 2.5. If the sinuosity of the river was 1.2 in Figure 16.2, then the adjusted sinuosity would be about 1.6 instead of 2.0. If the sinuosity of the river was 2.0 instead of 1.5 in Figure 16.3, then the adjusted sinuosity would be about 1.3 and instead of a straight channel, a sinuous channel would result.

Further evidence for the type of channel response shown in Figures 16.2 and 16.3 is demonstrated by the experimental studies of Jeff Ware (1992, personal communication). He lowered base-level relatively slowly to a maximum of 12 cm in a flume with a total length of 18.4 m. This change doubled the channel gradient. However, the effect of the base-level lowering extended only 4 m upstream, and the change in base-level was accommodated by an increase of sinuosity from 1.2 to 1.5 in the lower 4 m of the flume. It is clear, however, that pattern change did not totally compensate for the change of base-level. This channel widened and roughness increased, thereby assuming part of the adjustment to the base-level change.

The Manning Equation for velocity of flow (V) reveals that velocity is a function of hydraulic radius or average flow depth (R) stream gradient (S) and channel roughness (n):

$$V = \frac{1.49}{n} R^{2/3} S^{1/2} \qquad (16.2)$$

Ware's experiments showed that a sinuosity increase, which resulted in a slope decrease, was only part of the adjustment and both depth (R) and roughness (n) adjusted so as to decrease velocity and stream power. Therefore, pattern changes may only absorb part of a valley slope or base-level change, as suggested by Leopold and Bull (1979).

The discussion so far has been with regard to the response of sinuous streams to an increase or decrease of slope. A braided stream, however, may not be able to adjust as readily. Experimental studies

Figure 16.4 Longitudinal profiles of braided stream during baselevel lowering. Profiles indicate three stages of profile adjustment during the experiment (from Germanowski, 1989).

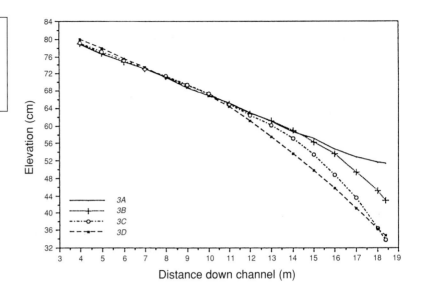

of the effect of base-level change on braided streams by Germanowski (1989) show that in these wide channels, that are formed in low cohesion sediments, a base-level lowering causes great channel instability. Nickpoints formed, as base-level was lowered at the end of a flume, but they quickly lost their identity by reclining. The amount of incision decreased progressively upstream until the rejuvenated reach merged with the unaffected upstream reach (Figure 16.4). After 8.5 hours, incision had not extended beyond 9 m upstream. With time, the channel would have undoubtedly adjusted further, but this experiment best illustrates the declining rate of incision upstream and the declining rate of sediment delivery, as shear stress is progressively distributed over a greater channel length. As noted earlier, the situation may be very different when a channel is confined, and its ability to shift laterally is restricted.

It can be concluded that, depending upon the circumstances of base-level lowering, there can be aggradation, degradation, or little change (Butcher, 1989). Nevertheless, this is a very simplistic view of the response of the fluvial system to change. For example, Figure 16.4 shows that, as incision took place in the lower part of the flume, there was aggradation upstream in response to the sediment delivery from upstream. When base-level was raised in this experiment, deposition eventually extended upstream to the 9 m limit of incision, but at the same time degradation was occurring above this point, as a result of the previous base-level lowering. In fact, in most incised channels, upstream degradation causes downstream aggradation so that the stream reaches are out of phase. Furthermore, studies of incised channels reveal that the increased sediment loads generated by incision will lead to deposition downstream (Figure 3.1). For example, stream capture rejuvenated a stream in Texas, which led to rejuvenation of tributary streams, larger sediment loads, and downstream deposition (Shepherd, 1979; Gardner, 1983). This response has also been noted during experimental rejuvenation of drainage net-works (Schumm *et al.*, 1987, p. 95). The end result of this complex response of rejuvenated fluvial systems is highly variable sediment delivery downstream, which is characteristic of fluvial sedimentary deposits (Schumm, 1981; Johnson *et al.*, 1985; Badgley and Tauxe, 1990; McRae, 1990) and channel reaches that are out of phase.

Chapter 17

Length

Very similar to the effect of base-level change is the effect of lengthening or shortening of a channel (Figure 1.2). This can occur by avulsion, meander cutoffs, and channelization. Shortening the channel will increase gradient and lengthening the channel will decrease gradient. A good example of this is meander change. A cutoff steepens gradient, and an increase of meander amplitude, as it grows, decreases gradient.

The braided rivers of the Canterbury Plains of New Zealand provide an example of fluvial response to increasing gradient caused by shortening of the channel (Leckie, 1994). The Canterbury Plains consist of coalescing alluvial fans, built during glacial episodes by the Rangitata, Ashburton, Rakaia, and Waimakariri rivers. All of these flow from the Southern Alps. Along the coast, the shoreline is retreating by about 1 m per year, and there are wave-cut cliffs up to 25 m high. The rivers are incising by up to 4 mm per year, with zones of incision extending 8–15 km inland. Upstream of this point, there is a zone of negligible incision, where the rivers flow at the original surface of the Plain.

A similar result occurs if a main channel shifts laterally across its floodplain thereby either lengthening or shortening a tributary channel. A lateral shift of a channel from one side of its floodplain to the other shortens streams on one side and lengthens them on the other. This causes degradation on one side and aggradation on the other, as gradients are steepened and made gentler.

An extreme example of the effect of shortening is that of the Homochitto River in southwestern Mississippi (Wilson, 1979). In the late 1930s, the Homochitto River upstream of its confluence with the Mississippi River was almost totally obstructed by trees, which had fallen into the channel. Flow was sluggish and overbank flooding frequent. To lessen flooding problems caused by the reduced channel conveyance of the lower reaches of the Homochitto River, the US Army Corps of Engineers began a major channel improvement project in the 1930s. On October 10, 1938, the Corps completed the Abernathy Channel, a cutoff running from the Homochitto River near Doloroso west to a point on the Mississippi River about 15 miles upstream of the

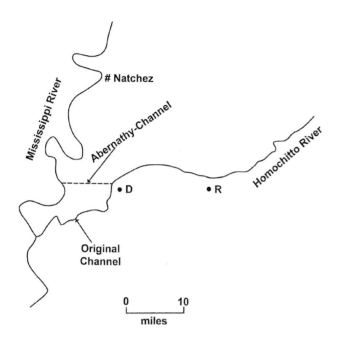

Figure 17.1 Map showing shortening of the Homochitto River in Mississippi by construction of the Abernathy channel. D and R indicate locations of Dolorosa and Rosetta (after Wilson, 1979).

mouth of the Homochitto River (Figure 17.1). This cutoff reduced the length of the river downstream from Doloroso from about 20 miles (*c.* 32 km) of meandering channel to a 9-mile (*c.* 14 km), relatively straight channel (Yodis and Kesel, 1993).

The 1938–1940 channel improvement project reduced the length of the channel resulting in an increase in the slope of the water surface and a reduction in channel resistance which increased stream velocities. The action of the higher flow velocities on the easily erodible sand and silt banks and streambed resulted in scour and channel degradation. This channel degradation probably began near the Mississippi River, and progressed upstream through the Abernathy Channel passing Doloroso by 1944. By 1949, the degradation of the streambed had progressed up the river past Rosetta and up several of its lower tributaries (Figure 17.1).

By 1944, significant degradation of the channel at Doloroso was noticeable, and by 1945 the channel had degraded 16.5 feet. The deepening of the channel weakened the foundation of the US Highway 61 bridge, and in 1950 substantial repairs were made to the bridge piers. During a flood in April 1955, the bridge collapsed, resulting in the loss of one life and three automobiles. Between 1957 and 1974 the channel at Doloroso had degraded about 19 feet (*c.* 579 m). This was accompanied by widening and general channel instability.

The Homochitto River example is an extreme example of river shortening by humans. Meander cutoffs also shorten a channel with results similar to that of the Homochitto River. Shortening and lengthening by meander cutoff and growth cause abrupt changes of channel character from meandering to braided and vice versa until the channel has adjusted.

Figure 17.2 Effect of channel lengthening across surfaces of different slope. See text for discussion.

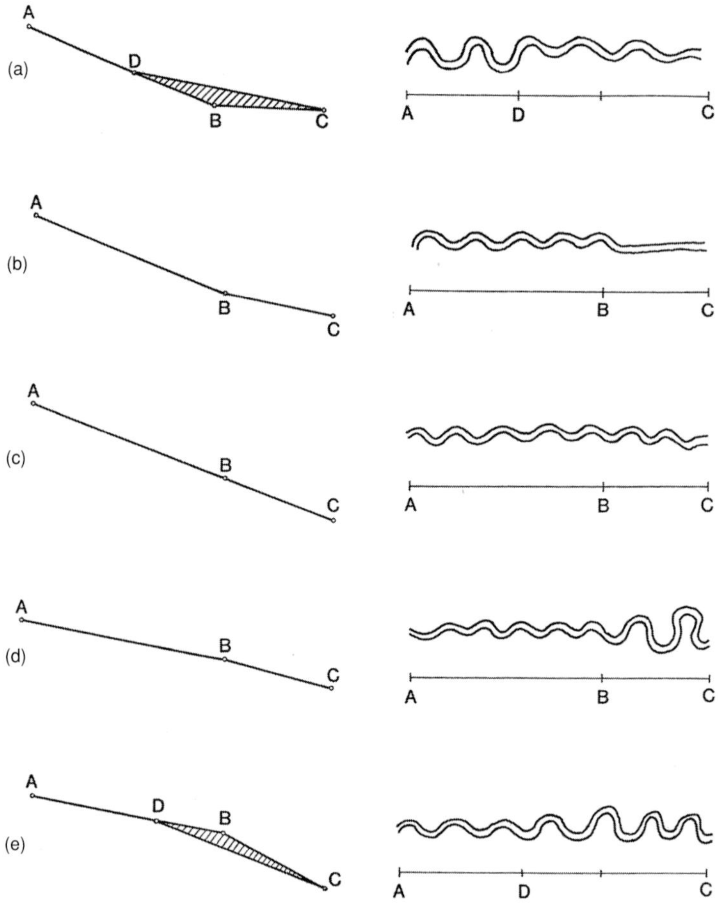

A prime example of the effect of lengthening is delta growth. As the Mississippi River delta prograded seaward, the channel of the Mississippi River lengthened and gradient decreased until the channel shortened by avulsion. These changes could dramatically alter channel morphology at least temporarily.

Figure 17.2 illustrates how a channel adjusts to lengthening over slopes of different inclination (BC) as a result of base-level change. It also illustrates channel adjustment to lengthening by pattern change. In examples (a) and (b) (Figure 17.2) the slope B C is gentler than that of the stream channel. In example (c), it is identical, and in examples (d) and (e) it is steeper. In example (a), the decrease in slope is large. The channel cannot adjust to this by a change of pattern or other channel characteristics so aggradation takes place. A wedge of sediment is deposited, which increases the gradient from B to C, but it decreases the gradient from D to C which is accommodated by a decrease of sinuosity, as in Figure 16.3. In example (b), the reduction of slope at B is not great, and the channel can compensate by straightening and reducing sinuosity without deposition. In example (c), the slope is identical to the slope of the stream, and therefore, the channel can extend itself across the surface without aggradation or degradation. Although as it does so, the new channel will probably form by developing natural levees and scouring slightly into the existing surface. In example (d), the slope is steeper than that of the stream, but the increase is not large, and the stream can accommodate this by an increase in sinuosity between B and C as in Figure 16.2. However, in example (e), the slope is much steeper, and the channel cannot adjust only by a pattern change. Incision takes place, and the channel probably widens dramatically and braids. The effect of this will be propagated upstream for some distance (C to D) until the channel can accommodate the slope change by a change in pattern, as well as by a change of shape and roughness (Equation 16.2). Where there has been a change of gradient, by base-level change or lengthening or shortening of the channel, a downstream morphologic channel change is assured and channel variability can be expected.

Part VI

Rivers and humans

The variability of rivers creates significant problems for humans, especially with regard to attempts to modify or control rivers. In addition, unintended consequences of human activities may have unfavorable results. Finally, very diverse rivers may determine the character of riparian civilizations.

Chapter 18

Applications

The previous discussions of river variability have applications to river management. The important concept is that river reaches vary in both location and through time. For example, an 11-mile (*c.* 17 km) reach of the Arkansas River (Figure 1.3) near Leadville, Colorado can be divided into eight reaches of different morphology. Gradient in this gravel-bed stream varies from 0.0067 to 0.011, sinuosity from 1.12 to 1.34, width from 60 to 110 feet (*c.* 18–33 m), calculated bankfull discharge from 330 to 1,060 cfs (*c.* 9–30 m³/s), and one reach is anastomosing. Any overall river modification or rehabilitation scheme would seem to fail because of the reach to reach variability of the river.

In order to prevent such problems, there are three concerns when undertaking practical work or, in fact, during any river investigation. These concerns are:

(1) An investigation should always consider not only the site of interest, but upstream and downstream river reaches to determine if the reach of concern is representative of the river. That is, an investigator should back away from the specific problem site and view it in a broader context.
(2) Rivers may range in sensitivity from very to not at all. An attempt should be made to evaluate river and reach sensitivity to determine if change is likely (Figure 11.2).
(3) The multiple hypothesis approach should always be considered in an attempt to explain or anticipate river behavior. That is, the most obvious conclusion may be incorrect.

Concern 1: a broader perspective

A good example of Concern 1 was the problems associated with a bridge over the Cimarron River (Figure 1.3) near Perkins, Oklahoma (Keeley, 1971). In 1953, a new bridge was constructed just downstream from the old Highway 177 bridge, which was judged to be in poor condition with erosion concentrated on the south bank about 1500 feet (*c.* 457 m) above the bridge abutment (Figure 18.1). In 1950, five pile

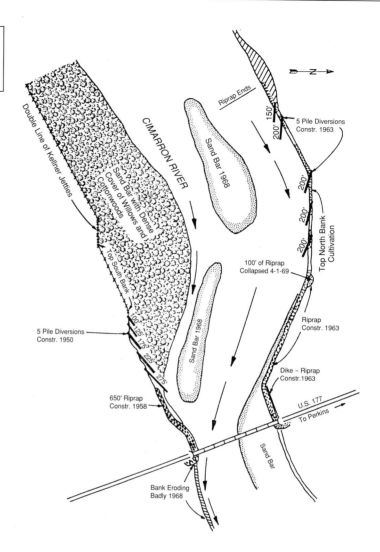

Figure 18.1 Map of Cimarron River near U.S. Highway 177, Perkins, Oklahoma (from Keeley, 1971).

diversion structures were constructed downstream from the old bridge site in order to protect the right (south) bank (Figure 18.1). In 1957, during a period of large floods, there was continued erosion of the south bank immediately upstream of the south abutment. Following the floods, 650 feet (*c.* 198 m) of riprap was emplaced on the south bank (Figure 18.1). In 1959, the second highest flood of record occurred and all five pile diversions were damaged. Also there was some bank erosion on the north bank 1500 feet (*c.* 457 m) upstream of north abutment.

Between 1959 and 1962, which was a period of high discharge, the point of attack shifted from the south bank to the north bank. Erosion occurred on the north bank for 2600 feet (*c.* 792 m) upstream of the north abutment. Five pile diversion structures were constructed on the north bank, and riprap was extended upstream from the north abutment (Figure 18.1). In 1965 there was further scour of north bank, and in 1969, 100 feet of riprap was lost by scour.

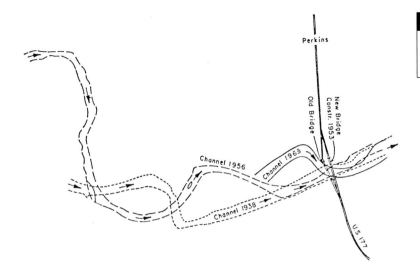

Figure 18.2 Cimarron River pattern in 1938, 1956, and 1968, near Perkins, Oklahoma (from Keeley, 1971).

In 1953 a cursory inspection of the channel upstream and a comparison of the situation with 1938 aerial photographs would have revealed that the major problem was meander shift (Figure 18.2). The continuing problem at this crossing could have been anticipated if a broader evaluation of the stability of the channel had been made prior to or after construction. Aerial photographs show that the channel was straight and braided at the site at the time of bridge construction, but there was a large meander about one mile (*c.* 1.6 km) upstream (Figure 18.2) and it was shifting downstream toward the bridge site.

As the bend approached the bridge, the cross-section changed from that of a relatively flat bed (a crossing) to a deeper channel (a pool). Relatively little effort would have been required to determine that the Cimarron River was a relatively unstable channel at this site and that a major problem was meander shift. In 1968 the apex of the meander had almost reached the bridge (Figure 18.2). Examination of the 1938 aerial photographs plus a rapid field examination of the channel would have revealed the problem.

The shift of the maximum erosion from the south to the north side of the bridge between 1959 and 1962 (Keeley, 1971) was the result of the movement of the bend into the area of the bridge (Figure 18.2). From the point of view of an engineer, the site selected in 1958 was a reasonable one, as the channel was straight, and it was near bedrock on the south side of the channel. Only if the upstream changes in the channel position were recognized could a highway engineer have anticipated the problems that developed at the new bridge site. According to Brice (1982), meander shift is one of the major problems at bridge crossings. Needless to say, this hazard should be one of the easiest to recognize if maps and aerial photographs for a period of years are available.

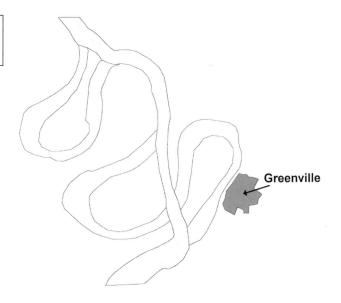

Figure 18.3 The Greenville Bends of the Mississippi River before and after cutoffs.

Greenville

Concern 2: sensitivity

An example of Concern 2 is that the deep arroyos of the southwestern USA may have formed as a result of sediment accumulation to a critical threshold slope that triggered incision during floods (Figures 12.5, 12.6). Another example of Concern 2 is the development of a meander to the point of cutoff. The probability of this can be evaluated by studying the shape of the meander. If a bend or bends looked like the Greenville Bends of the Mississippi River (Figure 18.3), cutoffs could be considered natural and imminent.

The work of Lane (1957) and Leopold and Wolman (1957) indicates that there is a gradient or discharge threshold above which rivers tend to be braided (Figures 18.4, 18.5). The experimental work reported by Schumm and Khan (1972) shows that for a given discharge, as valley floor slope is progressively increased, a straight river becomes sinuous and then eventually braids at high values of stream power and sediment transport (Figure 11.2). Rivers that are situated close to the meandering–braided threshold should have a history that is characterized by transitions in morphology from braided to meandering and vice versa.

The suggestion made here is that if one can identify the natural range of patterns along a river, then within that range the most appropriate channel pattern and sinuosity probably can be identified. If so, the engineer can work with the river to produce its most efficient or most stable channel. Obviously, a river can be forced into a straight configuration or it can be made more sinuous, but there is a limit to the changes that can be induced beyond which the channel cannot function without a radical morphologic adjustment. Identification of rivers that are near a pattern threshold would be useful, because a braided river near the threshold might be converted to a

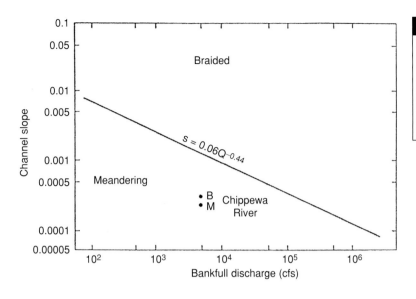

Figure 18.4 Leopold and Wolman's (1957) relation between channel patterns, channel gradient, and bankfull discharge. Letters B and M identify braided and meandering reaches of Chippewa River (Figure 1.3). cfs, cubic feet per second (0.0283 m³/s).

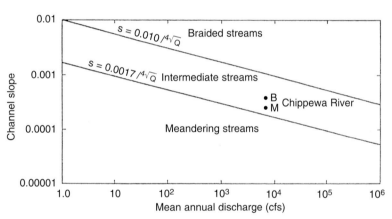

Figure 18.5 Lane's (1957) relation between channel patterns, channel gradient, and mean discharge. The regression lines were fitted to data from streams that Lane classified as highly meandering and braided. Between the two parallel lines are located streams of intermediate character ranging from meandering to braided. Letters B and M identify position of the braided and meandering reaches of the Chippewa River. cfs, cubic feet per second (0.0283 m³/s).

more stable single-thalweg stream (Figure 11.2). On the other hand, a meandering stream near the threshold should be identified in order that steps could be taken to prevent braiding, as a result of changes of land use and increased sediment load.

In most cases it will be difficult to determine if a river is susceptible to the type of treatment discussed in the preceding sections. Perhaps the best qualitative guide to river stability is a comparison of the morphology of numerous reaches, and the determination of whether or not there has been a change in the position and morphology of the channel during the last few centuries. Another approach might be to

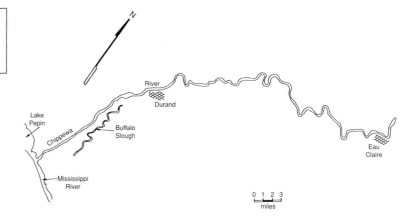

Figure 18.6 Map of the lower Chippewa River between Eau Claire, Wisconsin and Mississippi River.

determine the position of the river on the Leopold–Wolman (1957) or Lane (1957) gradient-discharge graphs (Figures 18.4, 18.5). If a braided river plots among the meandering channels or vice versa, it is a likely candidate for change because it is incipiently unstable.

An example is provided by the Chippewa River (Figure 1.3) of Wisconsin (Schumm and Beathard 1976), a major tributary to the Mississippi River (Figure 18.6). The Chippewa River rises in northern Wisconsin and flows 320 km to the Mississippi River. It is the second largest river in Wisconsin, with a drainage basin area of 24 600 km^2. From its confluence with the Mississippi to the town of Durand 26.5 km up valley, the Chippewa is braided (Table 18.1). The main channel is characteristically broad and shallow, and it contains shifting sand bars. The bankfull width is 333 m. The sinuosity of this reach is very low, being only 1.06. However, in the 68 km reach from Durand to Eau Claire, the Chippewa River has a meandering configuration with a bankfull width of 194 m and a sinuosity of 1.49. The valley slope and channel gradient are different for each reach of the river. The braided section has a gentler valley slope than the meandering reach upstream, 0.00035 as opposed to 0.00040, which is contrary to what is expected from Figure 11.2. However, the situation is reversed for channel slope. The braided reach has a channel gradient of 0.00033, whereas the meandering reach has a gradient of 0.00028.

At present there is no evidence to suggest that the Chippewa River is either progressively eroding or aggrading its channel. In fact, the river below Durand has remained braided during historic time. It has maintained its channel position and its pattern, but a significant narrowing, as the result of the attachment of islands and the filling of chute channels has occurred downstream of Durand, which resulted in a recent decrease in channel width of over 40 percent.

The relations described by Leopold and Wolman (1957) and Lane (1957) provide a means of evaluating the relative stability of the modern channel patterns of the Chippewa River. Bankfull discharge was plotted against channel slope on Figure 18.4 for both the braided and the meandering reaches of the Chippewa. The value used for the bankfull discharge is 53 082 cfs (c. 1502 m^3/s), which is the flood discharge

		Channel			Channel
Location	Channel pattern	width (m)	Sinuosity	Valley slope	slope
Below Durand	Braided	333	1.06	0.00035	0.00033
Above Durand	Meandering	194	1.49	0.0004	0.00028
Buffalo Slough	Meandering	212	1.28	0.00035	0.00027

Table 18.1 Chippewa river morphology

having a return period of 2.33 years. The braided reach plots higher than the meandering reach, but both are well within the meandering zone, as defined by Leopold and Wolman (1957). This suggests that the braided reach is anomalous; that is, according to this relation the lower Chippewa would be expected to display a meandering pattern rather than a braided one. Even when the 25-year flood of 98 416 cfs (c. 2785 m^3/s) is used, the braided reach still plots within the meandering region of Figure 18.4.

When the Chippewa data are plotted on Lane's graph (Figure 18.5) the same relation exists. The Chippewa falls in the intermediate region, but within the range of scatter about the regression line for meandering streams. Again the braided reach is seen to be anomalous because it should plot much closer to or above the braided stream regression line. The position of the braided reach, as plotted on both figures indicates that this reach should be meandering.

This conclusion requires an explanation that can be based on the geomorphic history of Chippewa River. For example, there is an as yet unmentioned significant morphologic feature on the Chippewa River floodplain. This is Buffalo Slough, which occupies the southeastern edge of the floodplain (Figure 18.6). It is a sinuous remnant of the Chippewa River that was abandoned, and it is evidence of a major channel change in the Chippewa River valley (Table 18.1).

Flow through Buffalo Slough has decreased during historic time, and indeed flow was completely eliminated in 1876, when the upstream end of Buffalo Slough was permanently blocked. The abandonment of the former Buffalo Slough channel by the Chippewa River is the result of an avulsion, but one that took many years to complete. The channel shifted from Buffalo Slough to a straighter, steeper course along the northwestern edge of the floodplain. This more efficient route gradually captured more and more of the total discharge. The new Chippewa channel was braided due to a higher flow velocity and the resulting bank erosion.

The sinuosity of Buffalo Slough is approximately 1.28 and the channel slope is about 0.00027. This channel slope is very similar to the channel slope of the meandering reach upstream of Durand, 0.00028, therefore, the meandering pattern of the Buffalo Slough channel was appropriate. However, this sinuous channel could not have transported the large amounts of sediment that the present braided channel carries to the Mississippi River, or it too would have followed a straight braided course. Therefore, the present sediment load carried by the Chippewa River is greater than that conveyed by the Buffalo Slough channel, as a result of the upper Chippewa River cutting into Pleistocene outwash terraces upstream. If the contribution of sediment from these sources were reduced, the lower Chippewa could resume its sinuous course.

An indication that the pattern conversion of the Chippewa could be successful if the upstream sediment sources were controlled is provided by the Rangitata River of New Zealand. The Rangitata River is the southernmost of the major rivers that traverse the Canterbury Plains of South Island. It leaves the mountains through a bedrock gorge. Above the gorge, the valley of the Rangitata is braided, and it appears that the Rangitata should be a braided stream below the gorge, as are all the other rivers which cross the Canterbury Plain. However, below the gorge, the Rangitata is meandering. A few miles farther downstream, the river cuts into high Pleistocene outwash terraces, and it abruptly converts from a meandering to a braided stream. The braided pattern persists to the sea. If the Rangitata could be isolated from the gravel terraces, it probably could be converted to a single-thalweg sinuous channel, because the Rangitata is obviously a river near the pattern threshold (Schumm, 1979).

Perhaps the simplest way to determine the most likely pattern for a river is to determine its pattern from the earliest maps and aerial photographs. If the river was braiding in the past, it seems unlikely that an attempt to convert to a sinuous channel will be successful. However, if the historic river was meandering and it is now braided, although there have been no major erosional or hydrologic changes in the drainage basin, then a conversion back to a meandering pattern is appropriate.

Concern 3: multiple thoughts

Concern 3 relates to the need to consider multiple hypotheses for prediction of landform behavior (Schumm, 1991). For example, if one has no information regarding a channel, it is not sensible to assume that the channel is not changing or will not change in the near future. Therefore, one should assume that the river could be stable and unchanging, active with normal change, or unstable with major morphological changes imminent or current. Obviously, investigation is required to determine which situation is correct.

An analogous situation relates to alluvial fans. As a young geomorphologist, the author was asked about hazards to pipelines on alluvial fans. His response was that an alluvial fan is a depositional landform and therefore, shallow burial would be appropriate. Unfortunately, the next question related to the presence of deep trenches on the fan. Obviously, the assumption that an alluvial fan is a broad cone over which flood waters spread widely in numerous shallow channels or as a sheet, is not always correct. In fact, alluvial fans vary widely in morphologic character (Figure 18.7).

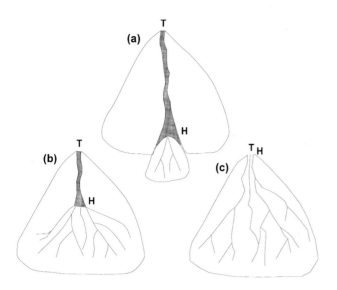

Figure 18.7 Three examples of alluvial-fan morphology. The letter T identifies the topographic apex which is the location where sediment and water from the upstream drainage basin enters the fan. The letter H identifies the location of the hydrographic apex, where channel flow becomes unconfined and produces alluvial-fan flooding.

The distribution of flood waters on a fan is critical for urban planning. If flood waters are spread uniformly over the surface of the fan, special efforts are required to protect structures. However, if the fan is incised, most of the fan is free of flooding or if it is partly incised, a large portion of the fan is free of flooding. Nevertheless, flood paths and the morphology of alluvial fans can differ greatly in space and time. For example, the sketches of Figure 18.7 show examples that form a continuum of alluvial fan types. Figure 18.7(a) shows a fan that has been trenched, and flow is confined to a single deep channel from the topographic apex (T) to the hydrographic apex (H), where the flow expands. On this type of fan, a highway crossing the toe of the fan is subject to alluvial-fan flooding, whereas a highway crossing the middle or upper part of the fan is affected only by changes of the incised channel. The greater part of this fan lies above the effects of flooding. Figure 18.7(b) shows a fan with a fan head trench which has incised the upper half of the fan. The hydrographic apex is at the downstream limit of the fan head trench. Most of this fan below the hydrographic apex is subject to flooding. Figure 18.7(c) shows a fan that does not have a well defined incised channel. The topographic and hydrographic apex occupy the same location, and most of the fan surface is subject to flooding.

This range of fan types has been described by Hunt and Mabey (1966) in Death Valley (Nevada) and observed through time during experimental studies (Schumm *et al.*, 1987). Therefore, within one area a range of fan types can occur, and during floods, fan morphology can change significantly.

A related situation along the Wasatch Mountains front in Utah emphasizes the need to consider alternative predictions and explanations of geomorphic phenomena. In May and June 1983, significant damage occurred along the Wasatch Mountains, Utah, due to snowmelt debris flows. The 1983 debris flows were triggered by landslides caused by a heavy snow pack and an abnormally late rapid snowmelt. Geologic studies of the structural fabric and hydrogeology of the land-slide source areas indicate these landslide-induced debris flows were a rare event. Most of the debris flows were generated by erosion during cloudburst storms which fell on watersheds depleted of vegetative cover by overgrazing and burning. Geologic studies of alluvial fans at the mouths of Wasatch Mountain canyons indicate that the majority of sediment incorporated into debris flows is derived from the drainage network. Surprisingly, this suggests that very high sediment delivery from the drainage basin may, in fact, be an indication of future stability. That is, stored sediment has been flushed from the drainage basin, and it may be a very long time before sufficient sediment accumulates again to produce debris flows even under extreme rainfall. This situation has been documented along the Watsach Mountain front north of Salt Lake City (Lowe, 1993; Keaton, 1995), where drainage basins that produced debris-flows in 1983 do not contain sufficient stored sediment to produce debris-flows at present. Therefore, not only the fan, but its drainage basin requires investigation in order to understand events in this region.

Approximately $12 million was spent to build or refurbish debris basins following the 1983 debris-flow events; less than $30,000 was spent on geologic research to understand the debris-flow processes. Had geologic studies been conducted prior to construction of the debris basins, more emphasis could have been placed on building debris basins at the mouths of canyons, which have not produced historical debris flows, instead of canyons that had produced recent debris flows. Clearly, the most likely canyons to produce large debris-flows in the near future are those which have not produced historical debris flows.

A final example of the need to consider alternatives is provided by the dredging that is necessary to maintain a navigation channel in the Mississippi River. It appears that there are problem reaches where frequent dredging is required. One suspects that hydraulic conditions at these locations cause the problem, but other controls may be functioning, and they should be considered. For example, at many of these locations, clay and coarse gravel have been sampled, which suggest some type of geologic control.

At river mile (RM) 899 near New Madrid (Figure 5.4), dredging is needed. Two miles (c. 3.2 km) downstream a fault crosses the river. The upthrown block is downstream of RM 899 and the clay sampled at this location probably acts as a submerged low dam. A similar situation exists at RM 885 where again dredging is required. A possible solution is to dredge a channel through the clay barrier, which would release the sand trapped upstream, and eliminate the need for dredging.

For more on techniques and applications of geomorphology to environmental management, see Thorne and Baghirathan, 1994; Thorne *et al.*, 1997; and Anthony *et al.*, 2001.

Chapter 19

Some unintended consequences

In Chapter 8, the consequences of human activities on rivers were discussed in passing. Here, some unintended consequences of river modification will be considered. For example, the meander cutoff program on the Mississippi River resulted in a straighter navigation channel but it also resulted in bank instability and the expenditure of vast amounts of money to stabilize the straighter, steeper channel. In addition, the changes of the Platte River, as its hydrology was altered (Chapter 7), and the increased flood stages of the Middle Mississippi River (Chapter 8) as the channel was confined by levees, are all examples of the unintended consequences of human activities.

Even the increase or decrease of gravel in a channel can significantly impact a river. Students of gravel bed rivers are well aware of the protective qualities of a gravel armor. It is perhaps less well recognized that small amounts of gravel in sand-bed rivers can have significant sediment transport and morphological effects. The Missouri and Nile rivers will be used to illustrate other impacts of small amounts of gravel on the morphology and response of sand-bed rivers.

For example, development of an armor significantly reduces bed-load transport in a Swiss river. This was demonstrated experimentally by Begin *et al.* (1980). In a large flume filled with sediment that contained only 1.2 percent of sediment larger than 2 mm, base-level was lowered in order to determine how sediment production changed with channel incision. Figure 19.1 shows a rapid decrease of sediment discharge from a peak as the channel reacted to the change of base-level. However, at about 300 hours into the experiment, a sharp decrease of sediment discharge occurred, as sediment larger than 2 mm armored the bed. The effectiveness of the armor occurred abruptly as recognized by Garde *et al.* (1977), and this is emphasized by the logarithmic plot of Begin's data (Figure 19.1).

Missouri River

The Missouri River (Figure 1.3) provides an example of how a small amount of gravel in the valley alluvium confounded the best plans

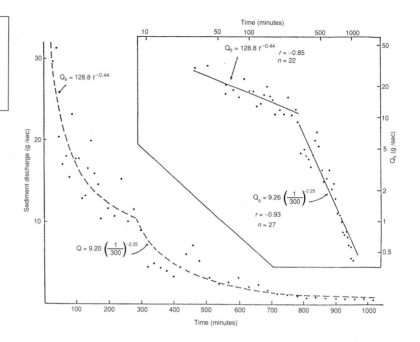

Figure 19.1 Changes with time of sediment discharge following base-level lowering in large flume. The same data are plotted arithmetically and logarithmically (from Begin *et al.*, 1980).

of hydraulic engineers, and it suggests how some climatic–hydrologic changes could, at least temporarily, influence river behavior.

Six large dams were constructed on the Missouri River between 1937 and 1963 (Figure 1.3). In preparation for construction of the Fort Randall Dam, estimates were made of the depth of scour that could be anticipated below the dam. This estimate depended to a large extent upon the size of sediment in the alluvium. Coring and bed samples contained sediment with a median size of 0.20 mm. Based upon this information, it was calculated that there would be 15 feet (*c.* 4.6 m) of degradation below the dam (Livesey, 1963). For the first two years after closure of the dam, water levels declined at a rate of 1 foot (0.3048 m) per year, but after 1954 the rate decreased to about one-tenth of that value (Figure 19.2). The reason for this decrease was revealed in 1962 when the bed of the river was exposed, during a period of no water release from the reservoir. The bed had armored (Figure 19.3). A very small amount of gravel in the alluvium had been concentrated on the bed during degradation, which greatly reduced the depth of erosion. The small amount of gravel in the alluvium was either not detected or ignored during sampling, and this gravel had a major impact on the river and dam operations. Immediately below the dam d_{90} increased from 0.35 mm to 10 mm. High-discharge releases from the dam apparently have been able to breach the armor, and 32 years after dam closure degradation has reached about 6.4 feet (*c.* 2 m).

In summary, a small amount of gravel, that was undetected during sampling, formed an armor that significantly inhibited degradation. It is also possible that the armor determined where degradation took place during high flows, and it may have shifted the tendency to erode

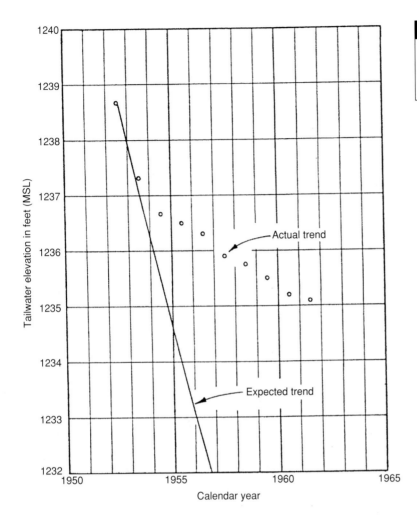

Figure 19.2 Change of water surface elevation following closure of Fort Randall Dam on the Missouri River (from Livesey, 1963). MSL, mean sea level.

Figure 19.3 Armor protecting underlying sand, Missouri River downstream of Fort Randall Dam (from Livesey, 1963).

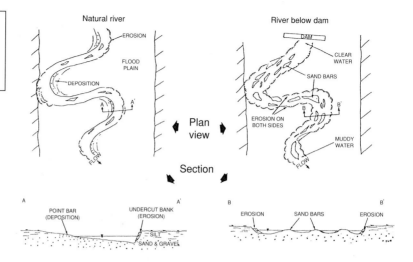

Figure 19.4 Idealized plan and cross-sections showing Missouri River characteristics before and after dam construction (from Rahn, 1977).

from the bed to the banks of the river. The river now erodes both banks on the inner and outer sides of meander and both banks at crossings (Rahn, 1977). The river appears to be replacing its sediment load by bank erosion, thereby changing the character of the river (Figure 19.4) by increasing its width–depth ratio. Xu (1996) found a similar response of a tributary to the Yangtze River when degradation below a dam exposed a gravel layer.

River Nile

The Nile is a relatively straight and stable sand-bed river (Schumm and Galay, 1994). Morphologic variability can be attributed to changes of valley slope, which causes adjustments of sinuosity and width. The valley slope changes reflect wadi contributions and geologic structures, such as faults. As in the Missouri River valley, the anticipated impact of the High Aswan Dam (HAD, see Figure 19.6) on hydrology and sediment loads was a matter of great concern for engineers involved with estimates of bank stability and potential channel degradation. Perhaps of greatest concern was the potential for major degradation of the Nile following construction of the dam. Potential degradation was studied by many researchers prior to (Fathy, 1956) and after construction of the HAD (Shalash, 1980; Shalash, 1983; Hammad, 1972). The construction of HAD commenced in 1963 and proceeded to 1968, and some degradation was expected from the HAD to the Esna Barrage (Figure 19.5). Measurements of water levels during the construction years downstream from each barrage showed some water-level lowering as far downstream as the Assiut Barrage, which is located about 600 km downstream (Figure 19.5). Pre-HAD estimates of degradation ranged from 2.0 to 8.5 m, but 18 years after the dam was in operation, maximum degradation below the barrages was only 0.70 m.

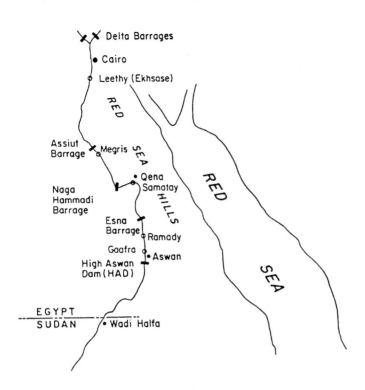

Figure 19.5 Index map of River Nile (from Schumm and Galay, 1994).

Several factors may account for the fact that degradation has been minimal after closure of HAD, but it is important to recognize that during past humid periods in Egypt, tributary wadis delivered coarse sediments to the valley, which could act as a control of degradation today, depending upon their location in the valley and the depth at which they are encountered. The most recent such period was between 11 000 and 6000 years ago (Paulissen and Vermeersch, 1987), but even today, wadi flooding must introduce coarse sediments into the river. For example, there is an abundant supply of sand, gravel, and cobbles in Wadi Qena, which drains a large area to the north of Qena (Figure 19.5). The bed of the large Wadi el Matuh, which enters the valley south of Quena appears very sandy, but gravel and some cobbles were observed on the alluvial surface. A deep trench was excavated in a small wadi that enters the Nile valley at Khuzam about 30 km upstream from Qena. Boulders and cobbles are abundant in the trench (Figure 19.6) and such sediments undoubtedly were and are moved into the Nile during wetter periods and during major floods. All of the wadis contain much stored sediment of gravel, cobble, and boulder size. Archaeologists and other investigators describe cobbles and gravel at wadi mouths (Wendorf *et al.*, 1970) and in older terrace deposits (Butzer, 1959).

Based upon an analysis of borings in the Nile valley, Attia (1954) concluded that within the valley, "coarse deposits composed of coarse sand, sand and gravel, or gravel which lie beneath the fine alluvial deposits, have an irregular upper surface. The depth from ground-surface to the top of these gravels, etc. varies from 8 to 26 m." This

Figure 19.6 Gravel and cobbles exposed in trench near mouth of Wadi Khuzam. Top of hat is 15 cm wide.

suggests that at many locations, the bed of the Nile could be in contact with sediment that is coarser than sand. Additional support for this hypothesis is provided by gravels encountered in the bores at Esna and Naga Hammadi barrages (Figure 19.5). At Naga Hammadi there was 25 percent gravel at a depth of 1 m in two bores. At Esna there was 9 percent gravel at the depth of 1 m in one bore. At other barrages gravels were encountered in the bores, but at greater depths. As shown by the Missouri River example and the Begin *et al.* (1980) experiment, an armor can develop from only a small percentage of coarse bed material.

There is a distinct change in both valley slope and river slope downstream of Qena. Coarse sediments introduced by Wadi Qena can resist erosion by the Nile, and they appear to act as a partial control of river slope. Although the bed sediment of the Nile ranges from 0.25 mm to 0.43 mm (Schumm and Galay, 1994) the coarse gravel and cobbles introduced by wadi discharges of the past and present probably produce gravel and cobble hard points in the bed of the river that only emerge from beneath the fine sand cover during floods when they prevent degradation.

The Missouri River and River Nile examples suggest that to understand river sedimentology and river behavior, a different sampling procedure may be required. Perhaps sampling needs to be done at high water when the fine sand is moving largely in suspension. Large boats and large samplers are needed in order to sample both the surface and subsurface sediment.

The above examples show the effect of degradation below dams, but the Little Tallahatchie River in northern Mississippi aggraded below Sardis Dam because the initial degradation lowered the local base-level for tributaries, which introduced large amounts of gravel into the river. The introduction of large amounts of gravel into the sand bed of the Little Tallahatchie River caused it to aggrade

(Biedenharn, 1983), and in order to maintain the channel, it was dredged below the dam.

Gravel stabilizes the bed, banks, and bars of the Mississippi River, and prevents excessive sediment movement. Gravel armored sand bars serve as semi-permanent channel controls. Removal of the gravel armor from the bars and islands can lead to erosion and loss of this control. As a result, meandering reaches may tend toward a braided character (Figure 19.4), and bed sediment transport may increase. Localized changes may even contribute to the deterioration of adjacent reaches. For example, removal of gravel makes it easier for the river to cut chute channels across point bars, which produce undesirable divided flows and cause erosion of previously protected middle bars that were semi-permanent islands in straight reaches.

The availability of a small percentage of gravel in the bed provides armoring material that may help to limit the depth and areal extent of scour around river contraction works such as spur dikes, and to limit degradation downstream from all types of hydraulic structures. These effects have a significant economic benefit in that the presence of gravel results in better protected river training works and structures. Thus, the uncontrolled removal of gravel can change the river, including its form, slope, velocity, transport capacity, stability, and environmental characteristics. Viewing the conclusions on the impact of gravel removal another way, it is clear that small amounts of gravel in alluvium can have a very significant role in river stability and channel morphology.

Two main conclusions can be stated as a result of the discussion of these three examples. First, small amounts of gravel can significantly control the morphology and behavior of sand-bed rivers. Secondly, sediment sampling of the bed during low flows may create a false impression of river sedimentology. A veneer of sand may obscure the important coarser bed sediment. Sampling should be done at least at medium discharges, and a robust sampler capable of collecting cobbles should be used.

Niobrara River

An impact similar to that of the Sardis Dam on the Little Tallahatchie River occurs at the junction of the Niobrara River (Figure 1.3) near Niobrara, Nebraska. The Niobrara River is a wide braided stream that delivers a load of well-sorted sand with a medium diameter of 0.012 mm. At the confluence with the Missouri, the Niobrara has formed a delta following construction of dams on the Missouri (Livesey, 1963). Upstream of the confluence, Fort Randall Dam has resulted in reduced mean and peak discharge, and Gavins Point Dam, downstream, has created a lake that extended to within 4.8 km of the mouth of the Niobrara by 1992. The competence of the Missouri River to transport sediment was greatly reduced (Livesey, 1963). As a result of these conditions, base-level (the level of the Missouri River) has

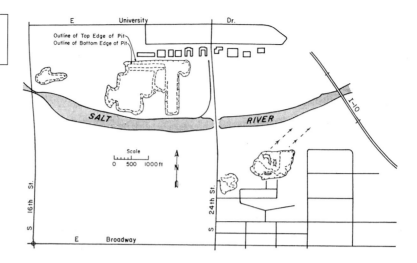

Figure 19.7 Location of Interstate (I-10) Bridge, Phoenix, Arizona and gravel pits.

risen and aggradation of up to 2.9 m has occurred at the mouth of the Niobrara River. Aggradation of the Niobrara River channel relative to the adjacent floodplains has created an alluvial ridge since initiation of the base-level rise. Surveys of channel cross-sections between 1956 and 1983 indicate that channel bed elevations have increased over 2 m, from a level 1.5 m below to a level 0.6 m above the adjacent floodplain.

Aggradation in the lower Niobrara River, as a direct result of the construction of Gavins Point Dam and flow regulation on the Missouri River, has extended approximately 20.9 km upstream from the confluence of the two rivers during the 43 years. The lower 3.3 km have experienced the most change, with major crevassing, avulsions, and a rising groundwater table, turning this portion of the river into an extensive wetland (Ethridge *et al.*, 1999). All of these changes have forced the movement of Niobrara State Park and the town of Niobrara onto the uplands adjacent to the Niobrara River Valley. In addition, a railroad bridge was abandoned and a highway bridge has been relocated and raised twice.

Salt River

A major problem for the transportation engineer is to anticipate upstream and downstream changes of floodplain and bank utilization and channel alterations. An example of what may be the worst possible case is provided by the Salt River at Phoenix, Arizona, where the river and its floodplain are a convenient and abundant supply of sand and gravel. The Salt River at Phoenix (Figure 19.7) is braided and sediment ranges from sand to boulders.

The Interstate-10 bridge over the Salt River was constructed in 1962. The bridge was designed to accommodate a 50-year flood with a peak discharge of 175 000 cfs (*c.* 4953 m³/s). Discharges were relatively low or nonexistent for a number of years, but a large flood

(67 000 cfs; *c.* 1896 m³/s) occurred in January 1966, and a 22 000 cfs (*c.* 623 m³/s) flood in April 1973. The river was essentially dry until a series of floods in 1978 and 1979. In March 1978, there was a 115 000 cfs (*c.* 3255 m³/s) flood, and it was followed by a 120 000 cfs (*c.* 3396 m³/s) flood in December 1978. In January 1979, there was an 80 000 cfs (*c.* 2264 m³/s) flood, and finally in March 1979, a 48 000 cfs (*c.* 1358 m³/s) flood.

During the last relatively small flood the bridge failed although it was designed to survive a 175 000 cfs (*c.* 4953 m³/s) flood. A 30-foot (*c.* 9 m) deep gravel pit had been excavated between S. 24th Street and the bridge (Figure 19.7). During the large floods, water flowed into the gravel pit and the upstream wall of the pit migrated toward the bridge until during the 48 000 cfs (*c.* 1358 m³/s) flood, the footings of the bridge were undermined and the bridge failed.

When the bridge was designed, it was assumed that the low water thalweg would remain fixed in position 5 feet (*c.* 1.5 m) above the deepest bridge pier. However, as the city of Phoenix grew during the period following bridge construction, gravel mining increased and gravel pits were opened near the bridge.

It is obvious that one cannot be too careful when modifying discharge and sediment loads of rivers. The small amount of gravel that inhibited degradation in the Missouri River could have been detected with a more careful sampling program, and the development of gravel controls in the Nile could have been anticipated if bed sediment sampling could have been undertaken at high discharges. In addition, removal of gravel from a channel can destabilize it.

Previous examples of unintended consequences are the conversion of an intermittent flow channel, from which a perennial flow channel allowed woody vegetation to greatly influence the morphology of the Platte River (Chapter 7). In addition, the constriction of the Middle Mississippi River and its effect on flood stages should have been readily anticipated, but it was not (Chapter 8).

The problems described in each case discussed are the result of concentrating too little on the broader situation, which would include the effect of downstream tributaries (Niobrara River), the detection of gravel at depth (Missouri and Nile Rivers), and hydrologic changes other than peak and mean annual discharge (Platte River).

Chapter 20

River impact on ancient civilizations: a hypothesis

Rivers comprise only a small part of a landscape, but much of the energy of the landscape is concentrated in the river channel, which often creates hazardous conditions for humans and vast floods along the world's great rivers create creation myths and misery. Rivers form boundaries between riparian properties, between states, and between nations. They shift and create international problems as along the Rio Grande border between the USA and Mexico. They provide routes for exploration, as in the opening of western US. The route down the Ohio, up the middle Mississippi, and up the Missouri let Louis and Clark finally reach the Pacific Ocean, after crossing the Bitterroot Mountains and continuing down the Snake and Columbia rivers. Exploration of Africa was primarily along the Nile, Niger, Congo, and Zambezi rivers.

Dwellers along great rivers could hardly not be influenced by behavior that ranged from relatively benign to dynamic. For example, river instability, as discussed in the previous chapters, could create environmental conditions leading to uncertainty for the riparian dweller. Diamond (1999) demonstrates the effect of environmental conditions on diverse human groups, and it may not be too problematic to suggest that river type may in the past have had major impacts on human perspectives. For example, if one river is stable and another subject to shift and avulsion, would not the perceptions of the two populations be different? For example, Macaulay in 1838 (Schama, 1995) commented on supposed affinities between French rivers and people. For example, the turbulent Garonne was the river of the impetuous Gascons which he contrasted with the Seine, which "flowing decorously" to the sea, was reflected in the "stalwart virtues of the Normans."

The emergence of the Sumerian, Egyptian and Harappan early civilizations occurred on the floodplains of the three alluvial valleys of the Tigris–Euphrates, the Nile, and the Indus. These valley floors offered an abundant water supply, fertile alluvial sediment, and of course, the possibility of irrigation. They also offered long distance water transport for travelers and goods. This was important as all riverine civilizations were poor in other resources. They were providers of abundant agriculture products, but for much else, trade and/or ready transport was necessary (Gibson, 1977). With so many things in common, nevertheless, the three civilizations emerged with diverse characteristics.

It is probably unwise to venture too far into a field that is not an area of one's expertise, especially if the effort is viewed as a resurrection of environmental determinism. Nevertheless, the field of environmental psychology deals with "interactions and relationships between people and their environments" (Proshansky, 1990). However, the emphasis has been on the reactions of individuals to the environment rather than large groups or societies (McAndrew, 1993). This chapter takes a giant step involving river behavior and the character of ancient civilizations.

The development of, and the nature and form of, great riverine (hydraulic) civilizations is a matter of considerable controversy (Wittfogel, 1957; Frankfort, 1959; Carneiro, 1970). Anthropologists, archaeologists and others have been especially concerned with the early Egyptian, Sumerian, and Harappan civilizations, that developed along the courses of the Nile, Euphrates, and Indus Rivers. The history and nature of these civilizations differ markedly, although all were undoubtedly subjected to the stresses of climatic and hydrologic fluctuations, invasions, wars, and plagues. The Egyptians maintained their civilization in the Nile valley with great success for thousands of years. The Sumerian civilization along the Euphrates River was characterized by the rise and fall of city states. The Harappans extended their civilization along the Indus River from the sea to the Himalayas. Much less is known about the Harappans, as their script has yet to be deciphered, but a major question is the reason for abandonment of their great cities, such as Mohen jo Daro and Harappa, when there is no evidence of warfare. Although oversimplified, the characteristics of these civilizations can be summarized as follows: Egyptian – long-term stability and continuity (Hawkes, 1973; Saggs, 1989); Sumerian – instability and flux (Gibson, 1973; Adams, 1981; Crawford, 1991); Harappan – stability followed by catastrophe (Dales, 1966; Lambrick, 1978).

The three rivers of interest here, the Nile, Indus and Euphrates all rise in areas of high precipitation and then flow into arid areas, which makes the inhabitants dependent upon their discharge for agriculture and their existence in a harsh environment. Nevertheless, in spite of this similarity the rivers are very different, just as the very stable Ohio River is in contrast to the unstable Missouri River, although both are major tributaries of the great Mississippi. In 1945 Mackay wrote a perceptive paper in which she attributed the abandonment of cities on the Euphrates and Indus plains to river avulsion. In contrast, the ancient Egyptian temples at Luxor and Kom Ombo and elsewhere still lie on the banks of the Nile after thousands of years.

As described earlier, major alluvial rivers differ in their morphology and dynamics in three ways. Because of these differences in location and time, the impact of different river morphology and behavior on ancient riverine civilizations should be significant. The differences can also be attributed to some extent to hydrologic factors (Saggs, 1989, p. 22). For example, the Nile floods in a very regular, predictable, and predominantly benign pattern between later summer and autumn after one harvest and before the next sowing. The Euphrates and Tigris are normally at their lowest in September and October and then begin to rise (Clawson *et al.*, 1971, p. 22). The Tigris is usually highest in March or April and the Euphrates a month later. At these times, flooding is likely to damage ripening crops. The Tigris flows faster and carries almost twice the amount of water of the Euphrates, making it difficult to control for irrigation. Consequently, most early major settlements in southern Mesopotamia were concentrated along the branching channels of the Euphrates (Saggs, 1989, p. 22, 23). In contrast, the Indus flow is low in November until midspring when snowmelt runoff from the Himalayas increases discharge. The highest discharge is in late July and August. The six-month period, May through October, accounts for 80 percent of the annual discharge (Milliman *et al.*, 1984, p. 66). Also, the Indus is subject to megafloods resulting from the failure of landslide and glacier ice dams in the Himalayas (Shroder, 1989). On average, the Egyptians were favored by more regular flows, but they too were subjected to years of low flow as well as to major floods.

The hypothesis advanced is that the differences among these civilizations and the abandonment of Harappan cities, is dependent upon the character of these great alluvial rivers. Each of the rivers is very different because of its geomorphic history, morphology, propensity for change, and type of change; therefore, geomorphic and geologic differences should be given serious consideration as the reason for the differences among these early civilizations. For example, the Nile is a river with a relatively gentle gradient which has shifted toward the east, but river behavior was a relatively minor problem for the Egyptians (Figure 20.1a). On the other hand, the steeper Indus channel has repeatedly avulsed, for long distances across the lower Indus Plain in Sindh (Figure 20.1b). The Euphrates was an anastomosing river with anabranches that diverge and rejoin (Figure 20.1c). Through time, the individual branches become inefficient as sinuosity increases, and they are abandoned. Towns located on these channels decline and eventually are abandoned, but at any one time some towns are declining while others are expanding, as more water is delivered through the channels that support them (Figure 20.1c).

Egyptians and the River Nile

The Egyptians were secure in their faith that all was right with the world. They lived on a stable river and were relatively secure from foreign attack (Saggs, 1989, p. 28). All Egyptian settlements were adjacent to one major river, and the narrowness of the Nile valley made it easy for a ruler to bring all of the Nile valley under his control

(a)

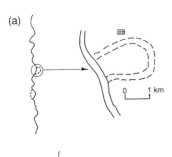

(b)

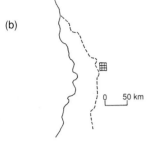

(c)

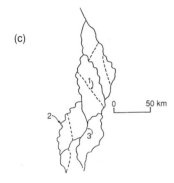

Figure 20.1 Types of avulsion. (a) Minor changes by meander cutoff and lateral shift, as in the Nile River Valley. (b) Major avulsion with development of a new channel some distance from its former course. A city on the old course will be abandoned, as on the Indus River Plain. (c) Avulsion within an anastomosing stream system, where channel segments are abandoned while others form, as on the Mesopotamian Plain. A city at location 1 will be abandoned, as the channels to the right and left capture its water. The city at location 2 is at its height, but the sinuous anabranch is becoming progressively inefficient, and its flow is being captured by the channel upon which city 3 is being established. City 2 will decline and city 3 will expand and become dominant.

(Saggs, 1989, p. 23). Competition for water was not an issue because whatever was done in any single natural floodbasin did not affect downstream water users (Butzer, 1976, p. 109).

Unlike most of the other great alluvial rivers of the world, the Nile is very straight (Figure 4.10). Between Qena and Cairo there are only a few short sinuous reaches. These occur on steeper reaches of the valley floor as a result of tributary contributions or faulting. In short, the River Nile is straight, gentle, and relatively stable.

Comparison of the early French maps of the Nile (Jacotin, 1826) with the modern river show minor change, although Butzer (1976, p. 34) identified 17 channel changes since 1600. Of these, 11 have been toward the east. In fact, there has been an eastward shift of some reaches of the Nile of as much as 3 km during the past 2000 years. On average, the shift is only 1.5 mm per year.

In summary, the relatively stable Egyptian civilization developed in the Nile valley along a river that was relatively stable. In Egypt, there was security and the ideal of changelessness in an eternal cycle (Hawkes, 1973, p. 22). Even if the river altered its position locally as demonstrated by Butzer, the effect was local and the overall river would not have been affected (Figure 20.1a). The stable and conservative Egyptian civilization was associated with a river that did not create major problems for the Egyptians, although great floods and years of low water would affect any riparian civilization.

The Sumerians and the Euphrates River

According to Hawkes (1973, p. 22), fundamental to the character of the Sumerians was a tendency to pessimism and insecurity. They believed that the gods created mankind to slave for them, and they lived in the expectation of disaster and return to chaos. The Mesopotamian plain was divided into city states characterized by rivalries that often flared into warfare and destruction, which is in keeping with this psychological character. According to Wenke (1990, p. 355) for centuries after 3000 BC, the Sumerian city states engaged in almost constant warfare with first one and then another gaining temporary ascendancy. Later, at about 2600 BC, almost all major settlements were fortified and there was conflict between the people of Ur, Uruk, Umma, and the other city states (Wenke, 1990, p. 338).

Channel avulsion on the Euphrates plain was a serious problem for towns, cities, and the individual farmer. For example, Adams (1958, p. 102) stated that the early settlements were arranged along the approximate position of ancient channels. However, utilization of particular channels was not continuous and major channels were abandoned. "To the ancient farmer, such a shift would have been as much of a disaster as a destructive flood . . . Rather than suggesting stability and continuity, the local river regime . . . must have seemed in flux and fraught with peril". In his many publications, Adams uses the term flux frequently (Adams, 1958, p. 102; 1973; pp. 79, 80; 1974, pp. 2, 3), and he refers to continual channel change as the situation on the Euphrates plain (Adams 1981, p. 160). He also recognized that the river system was anastomosing (Adams, 1973, p. 38; 1981, p. 7, 8). This is an important identification because a multiple-channel anastomosing or anabranching channel network is inherently unstable (Figure 20.1c).

A relatively recent example of anabranch development was described by Ionides (1937) using historical records. In 1837, the Euphrates flowed past the ruins of Babylon and Hillah to Somawah (Figure 20.2) with traces of older channels branching from the Euphrates near Museyib and passing near Nejef. Progressively more water left the Euphrates and flowed down this route. However, a through channel did not exist, and the marshes of the Bahr el Nejef formed (Figure 20.2). A channel carried water from the marshes at Shinafiyah. This channel in turn branched before reaching Somawah. By 1911 the Hindiyah was the main channel, and flow in the Hillah course was significantly reduced, but the upper and lower channels had not joined. Incision of the lower channel would have drained

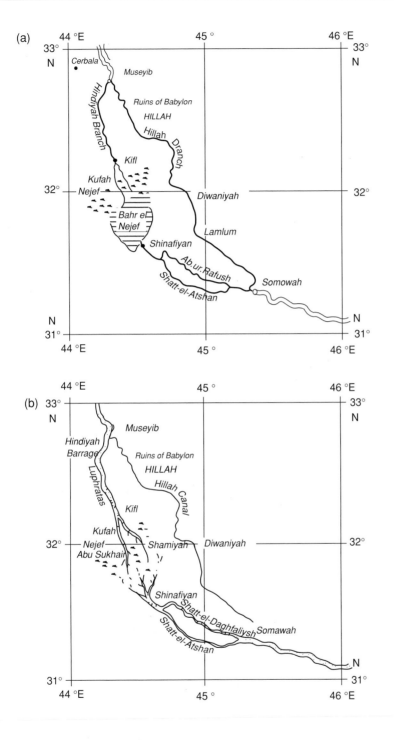

Figure 20.2 Anabranch formation, between Museyib and Somawah, Euphrates River: (a) condition of river in 1890; and (b) condition of river in 1937 (from Ionides, 1937).

the Bahr el Nejef and produced a through channel from Museyib to Somawah, but humans intervened to prevent draining of the Bahr el Nejef. This situation is identical to that on the Riverine Plain of southeastern Australia, where the Murray River and its tributaries the Ovens and King Rivers are anastomosing (Figure 3.7).

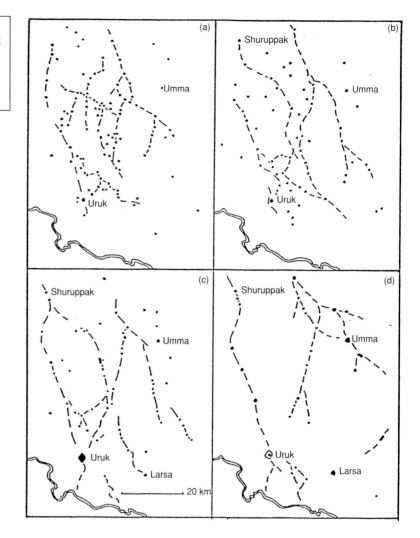

Figure 20.3 The Uruk countryside during: (a) Late Uruk; (b) Jamdet Nasr; (c) Early Dynastic I; and (d) Early Dynastic II/III periods (from Adams and Nissen, 1972 and Nissen, 1988).

It is obvious that loss of channel efficiency and avulsion will greatly affect each village, city, and individual farmer located on a decaying anabranch. In 1903 Cadoux (1906, p. 274) observed the effect of channel abandonment in an arid area. The Hillah Branch of the Euphrates between Musseyib and Somawah (Figure 20.2) became a dry channel, when the diversion structure (barrage) at Musseyib failed. Although efforts were being made to restore flow in the abandoned channel, villages along the river were nearly all abandoned. At the town of Hilla, holes were dug in the deepest part of the channel to supply water for the inhabitants. Water was encountered at an average depth of 3 feet (c. 1 m), but if water could not be reintroduced into the channel, the ground-water level would lower and eventually Hilla would be abandoned. In fact, Gibson (1973) developed a model for the rise of civilization (cities) that incorporates channel avulsion. He (Gibson, 1973, p. 454) concluded that abandonment of a major eastern channel of the Euphrates caused a population shift to the

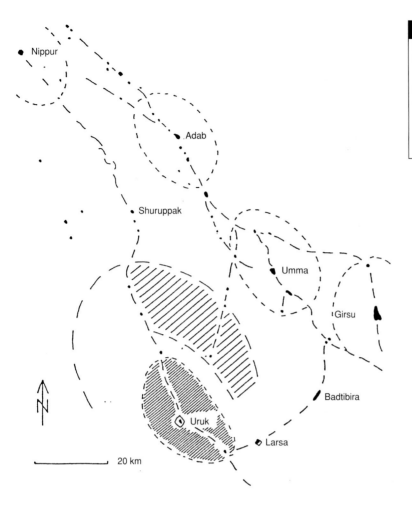

west and a concentration of population, which accelerated or even triggered the development of cities there.

Figure 20.3 shows how the pattern of anastomosing channels changed during thousands of years on the Mesopotamian Plain. Adams' extensive work in this area confirms that the channels form an anastomosing system of the Australian type (Figure 3.8). A map showing the zones of influence of some city states reveals how vulnerable they were to channel change (Figure 20.4). For example, if flow at Nippur were diverted into the Adab channel, the city of Uruk would be significantly affected. If the flow between Adab, and Umma were diverted toward Uruk before it reached Umma, both Umma and Girsu as well as Badtibira and Larsa would be affected. Conflict could hardly be avoided under these circumstances if one city interfered with downstream flow or if the inhabitants of a downstream city concluded that the natural flow decrease was the result of upstream interference.

In summary, the Sumerian civilization seemed destined to flux and conflict because of the nature of the river system. An

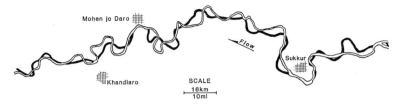

Figure 20.5 Indus River channel change between 1959 (black) and 1986 (white) near Mohen jo Daro (from Jorgensen *et al.*, 1993).

anastomosing network of channels is characterized by anabranches that form as overbank flood waters incise a new more efficient channel. Initially the higher velocity of flow and bank instability would not be favorable for navigation or settlement, but as the channel evolved to a condition of relative stability, it would support navigation and irrigation. However, with time, the channel would become more sinuous and less efficient. Deposition of sediment would further reduce the water carrying capacity and eventually the channel would be abandoned and the inhabitants forced to migrate. This situation is in great contrast to the stability of the River Nile.

Harappans and the Indus River

The Indus valley civilization is an enigma, and it is clearly different from the Egyptian and Sumerian civilizations. Much of the difficulty is that the Harappan script has not been deciphered, and therefore, there is no written description of the society (Miller, 1985). Without the insight provided by literature, the society has been described as static and monolithic or traditional (Miller, 1985). Also many Harappan sites do not lie along the Indus River, but rather are distributed widely in a variety of environments (Fairservis, 1971; Allchin and Allchin, 1982). Nevertheless, the great population centers of Harappan and Mohen jo Daro were located near major river channels, and the most heated discussion in the literature relates to the question of the abandonment of Mohen jo Daro.

The lower Indus River occupies a meanderbelt over which it migrates and floods (Figure 20.5). Hence, a city could not remain adjacent to the river, because of meander growth and shift. Indeed, the ruins of Mohen jo Daro are threatened, as a large bend of the river approached the site (Figure 20.5), but it is now protected by an embankment.

The very long occupation of Mohen jo Daro suggests river stability. However, it is clear that the Indus River has shifted tens and hundreds of kilometers across the Indus Plain. Avulsion on a grand scale has occurred and at times, more than one channel carried the flow from the Punjab to the sea (Figure 5.7). Therefore, many Harappan cities except Mohen jo Daro are located on abandoned channels (Allchin and Allchin, 1982), and these channels have been studied and discussed by many authors (Lambrick, 1964, 1967, 1973; Memon, 1969;

Holmes, 1968; Snelgrove, 1979; Flam, 1981, 1993). Numerous segments of ancient channels have been recognized and Flam (1981, 1993) has joined these remnants to show former Indus River positions.

Although the definition of avulsion is "an abrupt change in the course of a stream" (Bates and Jackson, 1987) this need not be instantaneous. In fact, the development of new distributary channels on the Mississippi River delta took hundreds of years, and the diversion of the Mississippi River down the Atachafalaya channel (Figure 1.3) was progressing slowly for at least a century until halted by engineering works (Fisk, 1944). As on the Euphrates plain, a new channel will slowly enlarge to accommodate the increasing discharge from the old channel until it has the capacity to carry the total flow of the old channel.

A progressive decrease of flow during the progress of an avulsion can cause the following: (1) less reliable water for irrigation and for a municipal water supply, which leads to crop failures and sanitation problems (Saggs, 1989, p. 121); (2) a reduction of water borne commerce and economic decline; (3) a fall of the ground water table necessitating deepening, of wells; and (4) a possible death of the riverine forests. As time passed, it would become progressively more difficult for the city to maintain contact with other areas and to feed itself.

In contrast to the multiple channels of Sumer, the Indus River was apparently a single channel that avulsed across its alluvial plain. The avulsion of this great river would have been a catastrophe for the Harappans, as not only would a city state be abandoned as in Sumer, but all of the riparian dwellers in cities, towns, and farms would lose their means of livelihood. Such a catastrophic event could cause the decline of a civilization.

The best example of such an event is the capture of the headwaters of the Nara Nadi or Hakra River by the Ganges River system and by a tributary of the Beas (Figure 20.6). Abandonment of numerous towns resulted (Wilhelmy, 1969; Mughal, 1982, 1990), including Ganweriwala, a mature Harappan city (85.5 ha.) that was larger than Harappa (65 ha.), but smaller than Mohen jo Daro (95.5 ha.).

It should be noted that at present, the Indus is flowing on the highest part of its valley. An avulsion would shift the channel to the east or west, and with time the ground water mound under the river would decline significantly causing failure of shallow wells that are now adjacent to the river.

The decline of Mohen jo Daro

The abandonment of Mohen jo Daro appears to be a great puzzle because the river is near the ancient city today (Figure 20.5), and as Lambrick (1973, p. 5) stated, "What the Harappans had the skill to do – and, we must add, the luck – was to select a site for their city as near perfect as possible; that is to say, near enough to the Indus or one of its great branches to have the advantage of flood irrigation in the

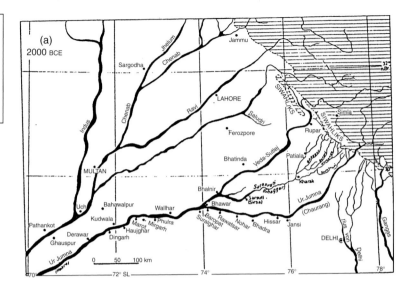

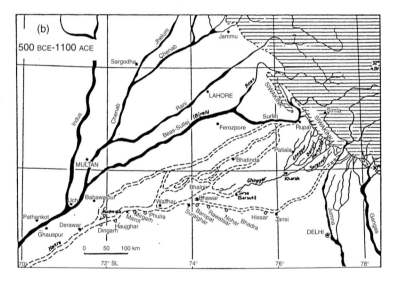

inundation season, without itself being invaded by these floods; not so near as to be in danger of erosion by the river's characteristic meander adjustments; and opposite a reach where it was unlikely to change its course appreciably." He further points out that "never, in a period approaching five thousand years, has the site been eroded by the Indus itself." Luck there may have been but skilled observation of flood occurrences probably led to the selection of the site or the city simply grew on a safe site. With regard to the modern Indus River the ruins are sited on a reach where flood heights are relatively low in comparison to flood elevations upstream and downstream (Harbor et al., 1994). This, however, could not prevent the infrequent major flood from causing extensive damage and perhaps temporary abandonment of the locality.

Table 20.1	Causes of abandonment
Human	Conquest, pillage and massacre
	Plague
	Exhaustion of timber resources
	Soil Salinization
Climatic and Hydrologic	Increased precipitation (increased flooding and longer duration of floods)
	Drought
	Megafloods (dam burst)
Tectonic	Earthquakes at the site
	Faulting at the site
	Faulting and damming of the river
Fluvial	Channel change (cutoffs, meandering)
	Avulsion

If the Indus River was not flowing past the site of Mohen jo Daro at present there would be little to argue about. For example, there is no doubt that Kalibangan, another Harappan city was abandoned when the river passing it lost its water supply by river capture (Raikes, 1968). Therefore, if a dry channel occupied the present Indus course no hypothesis other than avulsion would be required. Nevertheless, other factors may have been significant, and it is possible to think of about a dozen reasons for the abandonment of Mohen jo Daro that can be grouped into human, climatic, tectonic and fluvial causes (Table 20.1). Many of these can be eliminated because they would not lead to complete abandonment of the city or, if so, reoccupation at a later time would occur. For example, human causes such as conquest and plague would not cause total abandonment of valuable agricultural land, and although it has been proposed that the Harappans were conquered by Aryan invaders (Marshall, 1931; Mackay 1945; Piggott, 1953; Wheeler, 1966, 1968), with fewer than 20 bodies found within the city this does not suggest conquest and massacre. The description of large quantities of human bones in the ruins of Islamic Mansura to the west of the Indus River, which was taken and sacked, is very different from the situation at Mohen jo Daro (Cousens, 1929, p. 71). It is clear that the city of Mohen jo Daro was abandoned before the few unfortunates were killed in one part of the city (Dales, 1964).

There is a general consensus that the climate has not changed significantly since the establishment of the Harappan cities in the lower Indus valley, but climatic fluctuations, droughts and excessively wet years or decades do occur everywhere (Allchin and Allchin, 1982, p. 29, 32). Because the source of the Indus water is mainly from the Himalayas, drought on the lower Indus plain would not be significant for agriculture, but a significant decrease of Indus River discharge or a period of high, long-duration floods would cause temporary migration of the people to a source of food and safety. However, they certainly would return with a resumption of normal hydrologic conditions.

Another factor that could be of significance is that the Indus is subject to great floods produced by the failure of glacier and landslide dams in the mountains. When these discharges are superimposed on normal high water a catastrophic flood is the result (Hewitt, 1982; Burbank, 1983). Therefore, occasional megafloods undoubtedly created havoc for the Mohen jo Daro inhabitants. The mega floods could also have a great impact on the river itself, perhaps causing meander cutoffs and acting as a trigger of channel avulsion.

Although the Indus River flows through a tectonically active region, there is no mention in the archaeological literature of destruction of the city by earthquakes, although a city constructed of mud bricks and located on alluvium would be at great risk if a major earthquake occurred. Of course, no evidence of an active fault at the site has been presented, nevertheless, the effect of faulting elsewhere has led to a major controversy concerning the abandonment of the city. Raikes and Dales (1977) argue that the city was flooded by lake waters following fault damming of the river downstream, but no evidence for the dam exists (Lambrick, 1967; Possehl, 1967).

The remaining and most likely cause of the abandoning of Mohen jo Daro is fluvial (Table 20.1). There is no doubt that Mohen jo Daro was located relatively near a major Indus River channel. Flam (1981, p. 240) concludes that this river, which he named the Sindhu Nadi was located 25 km to the west of the city. The Sindhu Nadi was not functioning during Alexander the Great's conquest of Sindh in 326–315 BC (Lambrick, 1973, p. 88). The Sindhu Nadi lost its flow when the Ganges River system captured its headwaters (Figure 20.7). This caused abandonment of many towns including Kalibangan (Raikes, 1968; Stein, 1942; Oldham, 1893; Mughal, 1982, 1990). In contrast to the abandonment of a city by channel avulsion away from the site, Flam's detailed study of Indus River courses in Sindh leads him to conclude quite the opposite. Instead of a loss of water resulting from avulsion away from the city, he concludes that avulsion of a channel to a course close to the city subjected it to increased flooding, which led to its eventual abandonment (Flam, 1993). In any case, the long period involved in abandonment of the city and the migration of its inhabitants elsewhere (Gupta, 1982; Fairservis, 1967) fits the scenario of a slow change as avulsion progresses. Of course, such a major avulsion would affect many settlements along the old river course and cause a massive migration of the population.

As noted above, the presence of the Indus River just to the east of Mohen jo Daro is the chief reason for the controversy, and of course, its floodplain obscures any older course that was present. If a former course existed in this position, and then avulsed away from the city, the situation would be like that of other abandoned Harappan cities.

An objection to Flam's hypothesis is that it is unlikely for a river to abandon a course and then to occupy a nearby course or to reoccupy the old course. However, such a sequence of events has been recorded

in the Tigris–Euphrates valley by LeStrange (1905) where "in all probability as late as the middle of the 10th (16th) century, the Tigris when
it came about a hundred miles below Baghdad, turned off south, from
what is its present bed, flowing down by the channel now known as
the Shatt-al-Hayy (the Snake Stream) to Wasit . . . By this waterway
cargo boats went down without difficulty from Baghdad to Basrah . . ."
(p. 26). However, at present, the Tigris flows in its old eastern channel,
which it reoccupied sometime before 1581 (LeStrange, 1905, p. 28).
In 1652 the western channel still carried discharge but it was not
navigable.

In summary, channel avulsion as the primary cause of abandonment of Mohen jo Daro fits the known behavior of the river, as
recorded by the presence of numerous paleo-channels. Strong evidence for avulsion is the recorded position of channels during occupation of Mohen jo Daro and following its abandonment (Flam, 1993).
Circumstantial evidence is the slow decline of the city until it was
finally totally abandoned. This is consistent with decreasing discharge
and the decline of water levels in a nearby avulsing channel or as Flam
(1993) argued increasing flooding. The Tigris and Mississippi Rivers
provide examples of the avulsive behavior of large alluvial rivers such
as the Indus River. The location of the present course of the Indus
River is the fundamental cause of much of the controversy. If the river
had remained in a course away from Mohen jo Daro there would be
no controversy concerning the reasons for its abandonment as there
is none for Kalibangan and for numerous abandoned cities associated
with the Euphrates River.

Discussion

Fairservis (1971, p. 220) compares the characteristics of the three
ancient riverine civilizations: "Whereas Sumerian civilization is spec-
ulative, pessimistic in its world view, inventive and dynamic generally,
both Old Kingdom Egypt and Harappan India are striking in terms of
their conservative seemingly cultural form." For example, "The rapidity with which art styles change in Sumeria is almost opposite to
the stability of those styles in Egypt and in the Indus River Valley
civilization.

The stability of the Nile is reflected in the stability of the Egyptian
civilization. Furthermore, the Egyptians for the most part need not
expend energy repairing river damage to cities, and interruption to
other human activities. They had the time and energy to build great
temples and burial structures that were highly decorated. Although
the Harappans are linked to the Egyptians in Fairservis' comments
they were very different, and, indeed, the contrast of the Nile and
Indus is very great, and descriptions of both in 1843 are worth quot-
ing (Postans, 1843). "The Nile in its greatest size and volume falls very
far short of the magnitude of the Indus . . . The course of one river is
uniform and quiet, that of the other liable to sudden overwhelming

torrents." Postans observes that bank erosion along the Nile is minor whereas the Indus "constantly carries with its rapid current an immense portion of its banks; and thus the main stream is continually shifting and its bed undergoing complete alterations." In 1843 "The Nile, in every part of its stream which I have visited, is alive with boats of every size and description. This is, however, to be attributed to the populous country traversed by the latter, strikingly contrasted with the jungley and depopulated wastes through which the Indus flows, and where you may often journey for days, without seeing a sole or sign of human industry" (Postans, 1843).

Can the traditional character of the Harappan people be attributed to the fact that their energy was expended in annual flood fighting and cleanup? Did they spend their time in repairing flood damage and cleaning their drains instead of building pyramids, ziggurats, and creating art work? The 30 feet (c. 9 m) of flood deposits that bury the early stages of the city and upon which its ruins rest, support this concept. A more interesting question raised by most students regarding the Indus civilization was the reason for abandonment of its great cities. The explanations have been varied (Table 20.1), but in truth only one fits the facts. That is, channel avulsion and progressive loss of water supply, or as Flam (1993) concludes for Mohen jo Daro, avulsion and increased flooding. As channel avulsion progressed, abandonment of the city proceeded.

An avulsion, which involves abandonment of an old channel and formation of a new channel, is not instantaneous. The progressive diversion of water from the old channel would make river transport increasingly difficult. The new channel, which would be straighter and contain flows of higher velocity, would be difficult to navigate. In addition, bank failure as the channel widened would create hazards for boatmen. The result of an avulsion of the Indus type would affect large areas (Figure 20.1b), and, in fact, could cause the decline of the civilization as numerous towns are abandoned. Avulsion and anabranch formation of the Euphrates type would have a smaller effect (Figure 20.1c). Changes along the River Nile would be local and of short duration (Figure 20.1a).

The Sumerians coped with great competition among city states, which may have been the result of the anastomosing channel system of the Euphrates River. The difficulty and insecurity of their environment seems to have affected the form of Mesopotamian civilization causing a sense of insecurity, and an awareness of threat from careless gods (Hawkes, 1973, p. 23). There is an analogy in the Nile delta where a dozen little principalities existed on distributary channels (Hawkes, 1973, p. 27).

The hypothesis that river type affects the character of an early civilization appears to be reasonable. Obviously other factors are important and increased salinity of agricultural lands can be the cause of abandonment of agricultural areas (Jacobsen and Adams, 1958; Gibson, 1974, 1977, p. 1235). Nevertheless, the big alluvial rivers are important, and the stability and continuity of Egyptian civilization resembles the character of the River Nile. The instability and flux of Sumerian civilization resembles an anabranching river system. The lack of innovation in Harappan society may result from the struggle against the powerful, dynamic and avulsive Indus River. In fact, the decline of that civilization could have been precipitated by a major Indus River avulsion that affected all of the lower Indus valley in Sindh.

The hypothesis advanced here can only become a theory as more research is done, and if supporting evidence accumulates. It is advanced with the hope that it provides a new perspective for researchers in the three regions.

References

Abernethy, B. and Rutherfurd, I. (1998). Where along a river's length will vegetation most effectively stabilize streambanks? *Geomorphology* **23**: 55–75.

Adams, J. (1981a). Earthquakes, landslides and large dams in New Zealand. *NZ Natl. Soc. Earthquake Enginr.* **14**: 93–95.

(1981b). Earthquake-dammed lakes in New Zealand. *Geology* **9**: 215–219.

Adams, R. M. (1958). Survey of ancient water courses and settlement in central Iraq. *Sumer* **14**: 101–103.

(1973). Patterns of urbanization in early southern Mesopotamia. In C. Renfrew, ed., *The Explanation of Culture Change: Models in Prehistory*. London: Duckworth, pp. 735–749.

(1974). Historic patterns of Mesopotamian irrigation agriculture. In T. E. Downing and M. Gibson, eds., *Irrigations Impact on Society*. Anthropological Papers 25. Tucson: University of Arizona, pp. 1–6.

(1981). *Heartland of Cities*. Chicago: University of Chicago Press.

Adams, R. M. and Nissen, H. J. (1972). *The Uruk Countryside*. Chicago: University of Chicago Press.

Ahnert, F. (1970). Functional relationships between denudation, relief, and uplift in large mid-latitude drainage basins. *Am. J. Sci.* **268**: 243–263.

Albertson, P. E. and Patrick, D. M. (1996). Lower Mississippi River tributaries: contribution to the collective science concerning the "Father of Waters." *Engin. Geol.* **45**, 383–413.

Alford, J. J. (1982). San Vicente arroyo. *Assoc. Am. Geogrs. Ann.* **72**: 398–403.

Alford, J. J. and Holmes, J. C. (1985). Meander scars as evidence of major climate change in southwest Louisiana. *Assoc. Am. Geogrs. Ann.* **75**: 395–403.

Allchin, B. and Allchin, R. (1982). *The Rise of Civilization in India and Pakistan*. Cambridge: Cambridge University Press.

Anthony, D. J., Harvey, M. D., Laronne, J. B., and Mosley, M. P. (Eds.) (2001). *Applying Geomorphology to Environmental Management*. Highlands Ranch, CO: Water Resources Publications.

Attia, M. I. (1954). *Deposits in the Nile Valley and the Delta*. Cairo: Geological Survey of Egypt, Government Press.

Autin, W. J., Burns, S. F., Miller, R. T., Saucier, R. T., and Snead, J. I. (1991). Quaternary geology of the lower Mississippi valley. In R. B. Morrison, ed., *Quaternary Nonglacial Geology*. Conterminous U.S. (The Geology of North America Volume K-2). Geological Society of America, pp. 547–582.

Badgley, C. and Tauxe, L. (1990). Paleomagnetic stratigraphy and time in sediments: studies in alluvial Siwalik rocks of Pakistan. *J. Geol.* **98**: 457–477.

Baker, V. R., Kochel, R. C., and Patton, P. C. (Eds.) (1988). *Flood Geomorphology*. New York: Wiley.

Balling, R. C., Jr. and Wells, S. G. (1990). Historical rainfall patterns and arroyo activity within the Zuni River drainage basin, New Mexico. *Assoc. Am. Geogrs. Ann.* **80**: 603–617.

Bapat, A. (1988). Earthquakes and river regimes. In W. R. White, ed., *International Conference, River Regime*. Chichester: Wiley, pp. 423–429.

Barry, J. M. (1997). *Rising Tide.* New York: Simon and Schuster.

Bates, R. L. and Jackson, J. A. (Eds.) (1987). *Glossary of Geology.* Alexandria, VA: American Geological Institute.

Beck, S., Melfi, D. A., and Yalamenchili, K. (1984). Lateral migration of the Genesee River, New York. In C. M. Elliott, ed., *River Meandering.* New York: American Society of Civil Engineers, pp. 510–517.

Beeson, C. E. and Doyle, P. F. (1995). Comparison of bank erosion at vegetated and non-vegetated channel bends. *Water Resour. Bull.* **31**: 983–990.

Begin, Z. B. and Schumm, S. A. (1979). Instability of alluvial valley floors: a method for its assessment. *Trans. Am. Soc. Agric. Engs.* **22**: 347–350.

Begin, Z. B., Meyer, D. F., and Schumm, S. A. (1980). Nickpoint migration in alluvial channels due to baselevel lowering. *J. Waterways, Port, Coast. Ocean Enginr.* **106**: 369–388.

Begin, Z. B., Meyer, D. F., and Schumm, S. A. (1981). Development of longitudinal profiles of alluvial channels in response to base level lowering. *Earth Surf. Process. Landf.* **6**: 49–68.

Bell, F. H. (1999). *Geological Hazards.* London: Spon.

Belt, C. B. (1975). The 1973 flood and man's constriction of the Mississippi River. *Science* **189**: 681–684.

Biedenharn, D. S. (1983). Channel response of the Little Tallahatchie River downstream of Sardis Dam. In C. M. Elliott, ed., *River Meandering.* New York: American Society of Civil Engineers, pp. 500–509.

Bigler, W., Butler, D. R., and Dixon, R. W. (2001). Beaver-pond sequence morphology and sedimentation in northwestern Montana. *Phys. Geogr.* **22**: 531–540.

Blackbum, W. H., Knight, R. W., and Wood, M. K. (1982). *Impacts of Grazing on Watersheds.* Texas Agricultural Experimental Station, Texas A&H University College Station, MP 1496.

Blum, M. D. (1992). *Genesis and Architecture of Alluvial Stratigraphic Sequences: A Late Quaternary Example from the Colorado River, Gulf Coastal Plain of Texas.* Tulsa, OK: American Association of Petroleum Geology, Monograph No. 58, pp. 259–284.

Booth, D. B. (1991). Urbanization and the natural drainage system – impacts, solutions, and prognoses. *Northwest Environ. J.* **7**: 93–118.

Born, S. M. and Ritter, D. F. (1970). Modern terrace development near Pyramid Lake, Nevada and its geologic implications. *Geol. Soc. Am. Bull.* **81**: 1233–1242.

Boyd, K. F. and Schumm, S. A. (1995). *Geomorphic Evidence of Deformation in the Northern Part of the New Madrid Seismic Zone.* U.S. Geological Survey Professional Paper 1538-R.

Bradley, J. B. (1983). Transition of a meandering river to a braided system due to high sediment concentration flows. In C. M. Elliott, ed., *River Meandering.* New York: American Society of Civil Engineers, pp. 89–100.

Bray, D. I. and Kellerhals, R. (1979). Some Canadian examples of the response of rivers to man-made changes. In D. D. Rhodes and G. P. Williams, eds., *Adjustments of the Fluvial System.* Dubuque: Kendall-Hunt, pp. 351–372.

Brice, J. C. (1975). *Airphoto interpretation of the form and behaviour of alluvial rivers.* Final Report, US Army Research Office, Durham, Grant no. DA-ARD-D-31-124-70-G89.

(1981). *Stability of Relocated Stream Channels.* Washington, D.C.: Federal Highway Administration Report FHWA/RD-80/158.

(1982). *Stream Channel Stability Assessment.* Washington, D.C.: Federal Highway Administration, Report No. FHWA/RD-82/021.

(1983). Planform properties of meandering rivers. In C. M. Elliott, ed., *River Meandering.* New York: American Society of Civil Engineers, pp. 1–15.

Bristow, C. S. (1987). Brahmaputra River: channel migration and deposition. In F. G. Ethridge, R. M. Flores, and M. D. Harvey, eds., *Recent developments in Fluvial Sedimentology.* Society for Economic Paleontologists and Mineralogists, Special Publication 39, pp. 63–74.

Brizga, S. and Finlayson, B. (Eds.) (1999). *River Management, The Australian Experience.* Chichester: Wiley.

Brunsden, D., Jones, D. K. C., Martin, R. P., and Doornkamp, J. C. (1981). The geomorphological character of part of the low Himalaya of Eastern Nepal. *Zeit. Geomorph.* **37** (Suppl.): 25–72.

Brush, L. M., Jr. and Wolman, M. G. (1960). Nickpoint behavior in noncohesive material, a laboratory study. *Geol. Soc. Am. Bull.* **71**: 59–74.

Bryan, K. (1927). Channel erosion of the Rio Salado, Socorro County, New Mexico. *U.S. Geol. Surv. Bull.* **790**: 17–19.

(1928a). Change in plant associations by change in groundwater level. *Ecology,* **9**: 474–478.

(1928b). Historic evidence of changes in the channel of the Rio Puerco, a tributary of the Rio Grande, New Mexico. *J. Geol.* **36**: 265–282.

Bryant, M., Falk, P., and Paola, C. (1995). Experimental study of avulsion frequency and rate of deposition. *Geology* **23**: 365–368.

Bull, W. B. (1979). Threshold of critical power in streams. *Geol. Soc. Am. Bull.* **90**: 463–464.

(1991). *Geomorphic Responses to Climate Change.* Oxford: Oxford University Press.

Burbank, D. W. (1983). Multiple episodes of catastrophic flooding in the Peshawar basin during the past 700,00 years. *Geol. Bull. Univ. Peshawar* **16**: 43–49.

Burkard, M. B. and Kostachuk (1995). Initiation and evolution of gullies along the shoreline of Lake Huron. *Geomorphology* **14**: 211–219.

Burkham, D. E. (1972). *Channel Changes of the Gila River in Safford Valley, Arizona.* U.S. Geological Survey, Professional Paper No. 655-G.

Burnett, A. W. and Schumm, S. A. (1983). Alluvial river response to neotectonic deformation in Louisiana and Mississippi. *Science* **222**: 49–50.

Butcher, S. W. (1989). The nickpoint concept and its implications regarding onlap to the stratigraphic record. In T. A. Cross, ed., *Quantitative Dynamic Stratigraphy.* New York: Prentice Hall, pp. 375–385.

Butzer, K. W. (1959). Contributions to the Pleistocene geology of the Nile valley. *Erdkunke* **13**: 46–67.

(1976). *Early Hydraulic Civilization in Egypt.* Chicago: University of Chicago Press.

Butzer, K. W. (1980). Pleistocene history of the Nile Valley in Egypt and Lower Nubia. In M. A. J. Williams and H. Faure, eds., *The Sahara and the Nile.* Rotterdam: Balkema, pp. 253–280.

Cadoux, H. W. (1906). Recent changes in the course of the lower Euphrates. *Geogr. J.* **28**: 266–277.

Carneiro, R. L. (1970). A theory of the origin of the state. *Science* **169**: 733–738.

Carson, M. A. and LaPointe, M. F. (1983). The inherent asymmetry of river meander planform. *J. Geol.* **91**: 41–55.

Cecil, C. B., Dulong, F. T., Cobb, J. C., and Supardi (1993). *Allogenic and Autogenic Controls on Sedimentation in the Central Sumatra Basin as an*

Analogue for Pennsylvanian Coal-bearing Strata in the Appalachian Basin. Arizona: Geological Society of Amer., Special Paper No. 286, pp. 3–22.

Cenderelli, D. A. (2000). Floods from natural and artificial dam failures. In E. E. Wohl, ed., *Inland flood hazards.* Cambridge: Cambridge University Press, pp. 73–103.

Church, M. (1981). Reconstruction of the hydrologic and climatic conditions of past fluvial environment. In L. Starkel and J. B. Thornes, eds., *Paleohydrology of River Basins, British Geomorph.* Research Group, Technical Bulletin No. 28, pp. 50–107.

  (1983). Pattern of instability in a wandering gravel-bed river. In J. D. Collinson and J. Lewin, eds., *Modern and Ancient Fluvial Systems.* International Association Sedimentologists, Special Publication No. 6, pp. 169–180.

Church, M. and Gilbert, R. (1975). *Proglacial Fluvial and Lacustrine Sediments.* Society of Economic Paleontologists and Mineralogists, Special Publication No. 23, pp. 22–100.

Church, M. and Ryder, J. M. (1972). Paraglacial sedimentation: a consideration of fluvial processes conditioned by glaciation. *Geol. Soc. Am. Bull.* **83**: 3059–3072.

Clawson, M., Landsberg, H. H., and Alexander, L. T. (1971). *The Agricultural Potential of the Middle East.* New York: Elsevier.

Code, J. A. and Sirhindi, S. (1986). Engineering implications of impoundment of the upper Indus River, Pakistan, by an earthquake-induced landslide. In R. L. Schuster, ed., *Landslide Dams.* Geotechnical Special Publication No. 3. New York: American Society of Civil Engineers., pp. 97–110.

Coleman, J. M. (1969). Brahmaputra River channel processes and sedimentation. *Sed. Geol.* **3**: 129–239.

Collier, M., Webb, R. H., and Schmidt, J. C. (2000). *Dams and Rivers.* U.S. Geological Survey Circular No. 1126.

Cooke, R. U. and Reeves, R. W. (1976). Arroyos and environmental change in the American South-west. Oxford: Oxford University Press.

Costa, J. E. (1985). *Floods from Dam Failures.* U.S. Geological Survey Open-File Report No. 85-560.

Costa, J. E. and Schuster, R. L. (1988). The formation and failure of natural dams. *Geol. Soc. Am. Bull.* **100**: 1054–1068.

Cousens, H. (1929). The antiquities of Sind. *Archaeol. Surv. India* **46**: 1–184.

Crawford, H. (1991). *Sumer and the Sumerians.* Cambridge: Cambridge University Press.

Criss, R. E. (2001). Flood enhancement through flood control. *Geology* **29**: 875–878.

Dales, G. F. (1964). The mythical massacre at Mohen jo Daro. *Expedition* **6**: 36–43.

  (1965). New investigations at Mohen jo Daro. *Archaeology* **18**: 145–150.

  (1966). The decline of the Harappans. *Sci. Am.* **214**: 93–100.

Darby, S. E. and Simon, A. (Eds.) (1999). *Incised River Channels.* New York: Wiley.

Dedkov, A. P. and Moszherin, V. I. (1992). Erosion and sediment yield in mountain regions of the world. *IAHS Publ.* **209**: 29–36.

Denny, C. S. (1965). *Alluvial Fans in the Death Valley Region California and Nevada.* U.S. Geological Survey, Professional Paper No. 466: 1–62.

Desloges, J. R. and Church, M. (1987). Channel and floodplain facies in a wandering gravel-bed river. In F. G. Ethridge, R. M. Flores, and M. D. Harvey, eds., *Recent Developments in Fluvial Sedimentology.* Society for

Economic Paleontologists and Mineralogists, Special Publication No. 39: 99–109.

Diamond, J. (1999). *Guns, Germs, and Steel.* New York: Norton.

Dolan, R., Howard, A., and Trimble, D. (1978). Structural control of the rapids and pools for the Colorado River in the Grand Canyon. *Science* **202**: 629–631.

Dury, G. H. (1964). *Principles of Underfit Streams.* U.S. Geological Survey Professional Paper No. 452A.

Dynesius, M. and Nilsson, C. (1994). Fragmentation and flow regulation of river systems in the northern third of the world. *Science* **266**: 753–762.

Elliott, J. G., Gellis, A. C., and Aby, S. B. (1999). Evolution of arroyos: Incised channels of the southwestern United States. In S. E. Darby and A. Simon, eds., *Incised River Channels.* New York: Wiley, pp. 153–185.

Erskine, W. D. (1986a). River metamorphosis and environmental change in the Macdonald Valley, New South Wales, since 1949. *Aus. Geogr. Stud.* **24**: 88–107.

(1986b). River metamorphosis and environmental change in the Hunter Valley, New South Wales. Unpubl. Ph.D. dissertation, University of New South Wales.

Eschner, T. R., Hadley, R. F., and Crowley, K. D. (1983). *Hydrologic and Morphologic Changes in Channels of the Platte River Basin in Colorado, Wyoming, and Nebraska: A Historical Perspective.* U.S. Geological Survey Professional Paper No. 1277 A.

Ethridge, F. G., Skelly, R. L., and Brisco, C. S. (1999). *Avulsion and Meandering in the Sandy Braided, Niobrara River.* International Association of Sedimentology, Special Publication No. 28: 179–191.

Everitt, B. (1993). Channel responses to declining flow on the Rio Grande between Ft. Quitman and Presido, Texas. *Geomorphology* **6**: 225–242.

Fairbanks, R. G. (1989). A 17,000-year glacio-eustatic sea level record: influence of glacial melting rates on the Younger Dryas event and deep-ocean circulation. *Nature* **352**: 637–642.

Fairservis, W. A., Jr. (1967). *The Origin, Character, and Decline of an Early Civilization.* Novitiates, No. 2302. New York: American Museum Natural History, pp. 1–48.

Fairservis, W. A., Jr. (1971). *The Roots of Ancient India.* New York: MacMillan.

Fathy, A. (1956). *Some consideration on the degradation problem in the Aswan High Dam Scheme.* Alexandria, Egypt: University of Alexandria.

Fisk, H. N. (1943). Geological report on clay plugs in the Vicksburg Engineer District. Unpublished report Mississippi River Commission.

(1944). *Geological Investigation of the Alluvial Valley of the Lower Mississippi River.* Vicksburg, MS: Mississippi River Commission.

(1947). *Fine-Grained Alluvial Deposits and their Effects on Mississippi River Activity.* Vicksburg, MS: Waterways Experiment Station.

Flam, L. (1981). The paleogeography and prehistoric settlement patterns in Sind, Pakistan (*c.* 4,000–2,000 B.C.). Unpublished Ph.D. dissertation, University of Pennsylvania.

Flam, L. (1993). Fluvial geomorphology of the lower Indus Basin (Sindh, Pakistan) and the Indus civilization. In J. F., Shroder, Jr. ed., *Himalaya to the Sea.* London: Routledge, pp. 265–287.

Frankfort, H. (1959). *The Birth of Civilization in the Near East.* Bloomington: Indiana University Press.

Friedman, J. M., Osterkamp, W. R., Scott, M. L., and Auble, G. T. (1998). Downstream effects of dams on channel geometry and bottomland vegetation: Regional patterns in the Great Plains. *Wetlands* **18**: 619–633.

Froehlich, W., Kazowski, L., and Starkel, L. (1977). Studies of present-day and past river activity in the Polish Carpathians. In K. G. Gregory, ed., *River Channel Changes.* New York: Wiley Interscience, pp. 411–428.

Fuller, M. L. (1912). *The New Madrid Earthquake.* U.S. Geological Survey Bulletin No. 494.

Gagliano, S. M. and Thom, B. G. (1967). *Deweyville Terrace, Gulf and Atlantic Coasts.* Louisiana State University Coastal Studies Bulletin, No. 1, pp. 23–41.

Galay, V. J. (1983). Causes of river bed degradation. *Water Resour. Res.* **19**: 1057–1090.

Galay, V. (1987). *Erosion and Sedimentation in the Nepal Himalaya.* Ministry of Water Resources, Government of Nepal. Singapore: Kefford Press.

Garde, R. J. and Ranga Raju, K. G. (1977). *Mechanics of Sediment Transportation and Alluvial Stream Problems.* New York: Wiley.

Gardner, T. W. (1983). Experimental study of nickpoint and longitudinal profile evolution in cohesive, homogeneous material. *Geol. Soc. Am. Bull.* **94**: 664–672.

Gellis, A. C. (1996). Gullying at the Petroglyph National Monument, New Mexico. *J. Soil Water Conserv.* **51**: 155–159.

Gellis, A. C., Hereford, R., Schumm, S. A., and Hayes, B. R. (1991). Channel evolution and hydrologic variations in the Colorado River basin – Factors influencing sediment and salt loads. *J. Hydrol.* **124**: 317–344.

Germanowski, D. (1989). The effects of sediment load and gradient on braided river morphology. Unpublished Ph.D. dissertation, Colorado State University, Fort Collins.

Germanowski, D. and Schumm, S. A. (1993). Change in braided river morphology resulting from aggradation and degradation. *J. Geol.* **101**: 451–466.

Gibson, M. (1973). Population shift and the rise of Mesopotamian civilization. In C. Renfrew, ed., *The Explanation of Culture Change: Models in Prehistory.* London: Duckworth.

(1974). *Violation of Fallow and Engineered Disaster in Mesopotamian Civilization.* Anthropological Paper No. 24: University of Arizona, Tucson, pp. 7–19.

(1977). The breakdown of ancient desert civilization. *Desert Management,* vol. **4**. New York: Pengamon Press, pp. 1227–1237.

Gomez, B. and Marron, D. C. (1991). Neotectonic effects on sinuosity and channel migration, Belle Fourche River, western South Dakota. *Earth Surf. Proc., Landf.* **16**: 227–235.

Goodwin, C. N. (1999). Fluvial classification: Neanderthal necessity or needless normalcy. *Wildland Hydrol.,* **35**: 229–236.

Goudie, A. (1982). *The Human Impact.* Cambridge, MA: MIT Press.

Graf, W. L. (1983a). Flood-related channel change in an arid-region river. *Earth Surf. Proc., Landf.* **8**: 125–139.

(1983b). The arroyo problem – palaeohydrology and palaeohydraulics in the short term. In K. J. Gregory, ed., *Background for Palaeohydrology.* Chichester: Wiley, pp. 279–302.

(2001). Damage control: Restoring the physical integrity of America's rivers. *Ann. Assoc. Am. Geogr.* **91**: 1–27.

Gram, K. and Paola, C. (2001). Riparian vegetation controls on braided stream dynamics. *Water Resour. Res.* **37**: 3275–3284.

Gregory, K. J. (Ed.) (1977). *River Channel Changes.* Chichester: Wiley.

Gregory, K. J. and Brookes, A. (1983). Hydrogeomorphology downstream from bridges. *Applied Geogr.* **3**: 145–159.

Gupta, A. (1975). Stream characteristics in eastern Jamaica: an environment of seasonal flow and large floods. *Am. J. Sci.* **275**: 825–847.

Gupta, S. P. (1982). The late Harappan: a study in cultural dynamics. In G. L. Possehl, ed., *Harappan Civilization*. Oxford: New Delhi, pp. 51–59.

Gurnell, A. and Petts, G. (Eds.) (1995). *Changing River Channels*. Chichester: Wiley.

Hack, J. T. (1957). *Studies of Longitudinal Stream Profiles in Virginia and Maryland*. U.S. Geological Survey Professional Paper No. 294-B, pp. 45–97.

Hadley, R. F. and Schumm, S. A. (1961). *Sediment Sources and Drainage-Basin Characteristics in Upper Cheyenne River Basin*. U.S. Geological Survey Water-Supply Paper No. 1531-B, pp. 137–196.

Hall, S. A. (1990). Channel trenching and climatic change in the southern U.S. Great Plains. *Geology* **18**: 342–345.

Hammack, L. and Wohl, E. (1996). Debris-fan formation and rapid modification at Warm Springs Rapid, Yampa River, Colorado. *J. Geol.* **104**: 729–780.

Hammad, H. Y. (1972). River bed degradation after closure of large dams. *Hydraulics Div. Am. Soc. Civ. Engin.*, **98**: HY4.

Harbor, D. J., Schumm, S. A., and Harvey, M. D. (1994). Tectonic control of the Indus River in Sindh, Pakistan. In S. A. Schumm and B. R. Winkley, ed., *The Variability of Large Alluvial Rivers*. New York: American Society of Civil Engineers Press, pp. 161–175.

Harmon, M. E., Franklin, J. F., Swanson, F. J. *et al.* (1987). Ecology of coarse woody debris in temperate ecosystems. *Adv. Ecol. Res.*, **15**: 133–301.

Harvey, M. D., Pranger, H. S., Biedenharn, D. S., and Combs, P. (1988). Morphological and hydraulic adjustments of Red River from Shreveport, LA to Fulton, AK between 1886 and 1980. In S. R. Abt and J. Gassler, eds., *Hydraulic Engineering*. New York: American Society of Civil Engineers Press, pp. 764–769.

Harvey, M. D. and Schumm, S. A. (1987). Response of Dry Creek, California, to land use change, gravel mining and dam closure. *Int. Assoc. Hydrol. Sci. Pub.*, **165**: 451–460.

(1994). Alabama River: Variability of overbank flooding and deposition. In S. A. Schumm and B. R. Winkley, eds., *The Variability of Large Alluvial Rivers*. New York: American Society of Civil Engineers Press, pp. 313–337.

Hawkes, J. (1973). *The First Great Civilizations*. New York: Knopf.

Heede, B. H., Harvey, M. D. and Laird, J. R. (1988). Sediment delivery linkages in a chaparral watershed following a wildfire. *Environ. Manage.* **12**: 349–358.

Heller, P. L. and Paola, C. (1996). Downstream changes in alluvial architecture: an exploration of controls on channel-stacking patterns, *J. Sed. Res.* **66**: 197–306.

Hereford, R. (1993). *Entrenchment and Widening of the Upper San Pedro River*. Arizona: Geological Society of America Special Paper No. 282.

Heritage, G. L., Charlton, M. E., and O'Regan, S. (2001). Morphological classification of fluvial environments: an investigation of the continuum of channel types. *J. Geol.* **109**: 21–33.

Hession, W. C., Pizzuto, J. E., Johnson, T. E., and Horwitz, R. J. (2003). Influence of bank vegetation on channel morphology in rural and urban watersheds. *Geology* **31**: 147–150.

Hewitt, K. (1982). *Natural Dams and Outburst Floods of the Karakoram Himalaya.* Internat. Assoc. Hydrologic Science Publication No. 138, p. 259–269.
  (1998). Catastrophic landslides and their effects on the upper Indus streams, Karakoram Himalaya, northern Pakistan. *Geomorphology* **26**: 47–80.
Hey, R. D. and Thorne, C. R. (1986). Stable channels with mobile gravel beds. *J. Hydraulic Eng.* **118**: 671–689.
Hicks, D. M. and Davies, T. (1997). Erosion and sedimentation in extreme events. In M. P. Mosley and C. P. Pearson, eds., *Floods and Droughts, the New Zealand Experience.* Christchurch: New Zealand Hydrological Society, pp. 117–142.
Hicks, D. M., Hill, J., and Shankar, U. (1996). Variation of suspended sediment yields around New Zealand: the relative importance of rainfall and geology. In D. E. Walling and B. W. Webb, eds., *Erosion and Sediment Yield: Global and Regional Perspectives.* IAHS Publication No. 236, pp. 149–156.
Higgins, C. G. (1990). *Gully Development.* Arizona: Geological Society of America Special Paper No. 252, pp. 139–156.
Holbrook, J. and Schumm, S. A. (1999). Geomorphic and sedimentary response of rivers to tectonic deformation. *Tectonophysics* **305**: 287–306.
Holland, W. N. and Pickup, G. (1976). Flume study of nickpoint development in stratified sediment. *Geol. Soc. Am. Bull.* **87**: 76–82.
Holmes, D. A. (1968). The recent history of the Indus. *Geogr. J.* **134**: 367–382.
Hovius, N. (1998). Controls on sediment supply by large rivers. In K. W. Shanley and P. J. McCabe, eds., *Relative Role of Eustatic, Climate and Tectonics in Continental Rocks.* Society of Economic Paleontologists and Mineralogists Special Publication No. 59, pp. 3–16.
Hovius, N., Stark, C. P., and Allen, P. A. (1997). Sediment flux from a mountain belt derived by landslide mapping. *Geology* **25**: 231–234.
Hsü, K. J. (1983). *The Mediterranean was a desert: A voyage of the Glomar Challenger.* Princeton, NJ: Princeton University Press.
Huang, H. Q. and Nanson, G. C. (1997). Vegetation and channel variation: A case study of four small streams in southeastern Australia. *Geomorphology* **18**: 237–249.
Huckleberry, G. (1994). Contrasting channel response to floods on the middle Gila River, Arizona. *Geology* **22**: 1083–1086.
Humphreys, A. A. and Abbot, H. L. (1876). *Report on the Physics and Hydraulics of the Mississippi River.* Washington, D.C.: U.S. Government Printing Office.
Hunt, C. B. and Mabey, D. R. (1966). *The Stratigraphy and Structure, Death Valley, California.* U.S. Geological Survey Professional Paper No. 494-A.
Hupp, C. R. (1999). Relations among riparian vegetation, channel incision processes and forms, and large woody debris. In S. E. Darby and A. Simon, eds., *Incised River Channels.* New York: Wiley, pp. 219–245.
Ikeda, S. and Izumi, N. (1990). Width and depth of self formed straight gravel-bed rivers with bank vegetation. *Water Resour. Res.* **26**: 2353–2364.
Ionides, M. G. (1937). *The Regime of the Rivers Euphrates and Tigris.* London: Spon.
Jacobsen, T. and Adams, R. M. (1958). Salt and silt in ancient Mesopotamian agriculture. *Science* **128**: 1251–1258.
Jacotin, M. (1826). *Description de l' Egypte Carte topographique de l' Egypte (1:100,000),* 2nd edn., Paris.

James, L. A. (1991). Incision and morphologic evolution of an alluvial channel recovering from hydraulic mining sediment. *Geol. Soc. Am. Bull.* **103**: 723–736.

Jansson, M. B. (1988). A global survey of sediment yield. *Geogr. Ann.* **70**A: 81–98.

Jessop, W. (1782). *Report of William Jessop, Engineer, on a Survey of the River Trent in the Months of August and September, 1782, Relative to a Scheme for Improving its Navigation.* Nottingham: Burbage & Son.

Johnson, N. M., Stix, J., Tauxe, L., Cerveny, P. F., and Tahirkeli, R. A. K. (1985). Paleomagnetic chronology, fluvial processes, and tectonic implications of the Siwalik deposits near Chinji Village, Pakistan. *J. Geol.* **93**: 27–39.

Johnson, S. Y. and Alan, A. M. N. (1991). Sedimentation and tectonics of the Sylheth trough, Bangladesh. *Geol. Soc. Am. Bull.* **103**: 1513–1527.

Johnson, W. C. (1994). Woodland expansion in the Platte River, Nebraska: patterns and causes. *Ecol. Monogr.* **64**: 45–84.

(1998). Adjustment of riparian vegetation to river regulation in the Great Plains, USA. *Wetlands.* **18**: 608–618.

Jones, J. A. A. (1997). Subsurface flow and subsurface erosion. In D. R. Stoddart, ed., *Process and Form in Geomorphology.* London: Routledge, pp. 74–120.

Jones, L. S. and Blakey, R. C. (1997). Eolian-fluvial interaction in the Page Sandstone (Middle Jurassic) south-central Utah – a case study of erg-margin processes. *Sediment. Geol.* **109**: 181–198.

Jones, L. S. and Harper, J. (1998). Channel avulsions and related processes, and large-scale sedimentation patterns since 1875, Rio Grande, San Luis Valley, Colorado. *Geol. Soc. Am. Bull.* **110**: 411–421.

Jones, L. S. and Schumm, S. A. (1999). Causes of avulsion: An overview. In N. D. Smith and J. Rogers, eds., *Fluvial Sedimentology V1*, International Association of Sedimentology, Special Publication 28, pp. 171–178.

Jorgensen, D. J., Harvey, M. D., Schumm, S. A., and Flan, L. (1993). Morphology and dynamics of the Indus River: implications for the Mohen jo Daro site. In J. F. Shroder, Jr., ed., *Himalaya to the Sea: Geology, Geomorphology and the Quaternary.* London: Routledge, pp. 288–326.

Kazmi, A. H. (1979). Active faults in Pakistan. Geodynamics of Pakistan. In A. Farah and K. A. DeJong, eds., *Quetta: Geological Survey of Pakistan.*

Keaton, J. R. (1995). Dilemmas in regulating debris-flow hazards in Davis County, Utah. In W. R. Lund, ed., *Environmental and Engineering Geology of the Wasatch Front Region.* Utah Geological Association Publication No. 24, pp. 185–192.

Keeley, J. W. (1971). *Bank Protection and River Control in Oklahoma.* Oklahoma: Federal Highway Administration, Oklahoma Division.

Keller, E. A. and Pinter, N. (1996). *Active Tectonics.* Upper Saddle River, NJ: Prentice-Hall.

Keller, E. L. and Swanson, F. J. (1979). Effects of large organic material on channel form and fluvial processes. *Earth Surf. Proc. Landf.* **4**: 361–380.

Kellerhals, R., Church, M., and Bray, D. I. (1976). Classification and analysis of river processes. *J. Hydraul. Div. Proc.* **102**: 813–829.

Kennedy, R. G. (1895). The prevention of silting in irrigation canals. *Minutes Proc. Civil Engs.* **119**.

Kesel, R. H. and Yodis, E. G. (1992). Some effects of human modifications on sand-bed channels in southwestern Mississippi, USA. *Environ. Geol. Water Sci.* **20**: 93–104.

Kircher, J. E. and Karlinger, M. R. (1983). *Effects of Water Development on Surface-Water Hydrology, Platte River Basin in Colorado, Wyoming, and Nebraska Upstream from Duncan, Nebraska*. U.S. Geological Survey Professional Paper No. 1277-B.

Knighton, A. D. (1987). River channel adjustment – the downstream dimension. In K. Richards, ed., *River Channels*. Oxford: Blackwell, pp. 95–128.

Knighton, A. D. (1989). River adjustment to changes in sediment load: the effects of tin-mining on the Ringarooma River, Tasmania, 1875–1984. *Earth Surf. Proc. Landf.* **14**: 333–359.

Knighton, A. D. and Nanson. (1993). Anastomosis and the continuum of channel pattern. *Earth Surf. Proc. Landf.* **18**: 613–625.

Knox, J. C. (1983). Responses of river systems to Holocene climates. In H. E. Wright, Jr., ed., *Late Quaternary Environments of the United States, The Holocene*. Minneapolis: University of Minnesota Press, pp. 26–41.

Kochel, R. C. (1988). Geomorphic impact of large floods: review and new perspectives on magnitude and frequency. In V. R. Baker, R. C. Kochel and P. C. Patton, eds., *Flood Geomorphology*. New York: Wiley, pp. 169–187.

Kozarski, S. and Rotnicki, K. (1977). Valley floors and changes in river channel patterns in the north Polish Plain during the lake Würm and Holocene. *Quaest. Geogr.* **4**: 51–93.

Krinitsky, E. L. (1965). *Geological Influences on Bank Erosion along Meanders of the Lower Mississippi River*. Waterways Experiment Station, Vicksburg, MS: Potamology Investigation Report No. 12–15.

Lacey, G. (1930). Stable channels in alluvium. *Inst. Civil Engin. Proc.* **229**: 259–292.

Lagasse, P. F. (1986). River response to dredging. *J. Waterway, Port, Coast. Ocean Enginr.* **112**: 1–14.

Laird, J. R. and Harvey, M. D. (1986). Complex response of a chaparral drainage basin to fire. *Int. Assoc. Hydrol. Sci. Pub.* **159**: 165–183.

LaMarche, V. C., Jr. (1966). *An 800-year History of Stream Erosion as Indicated by Botanical Evidence*. U.S. Geological Survey Professional Paper No. 550-D, pp. 83–86.

Lambrick, H. T. (1964). *Sind, A General Introduction*. Hyderabad: Sindhi Adabi Board.

(1967). The Indus floodplain and the Indus civilization. *Geogr. J.* **133**: 483–495.

(1973). *Sind before the Muslim Conquest*. Hyderabad: Sindhi Adabi Board.

Lane, E. W. (1937). Stable channels in erodible materials. *Am. Soc. Civil Engin.* **102**: 123–194.

(1955). The importance of fluvial morphology in hydraulic engineering. *Am. Soc. Civil Engin. Proc.* **81** (745): 1–17.

(1957). *A Study of the Shape of Channels Formed by Natural Streams Flowing in Erodible Material*. Omaha, NE: U.S. Army Engineer Division. M.R.D. Sediment Series No. 9.

Langbein, W. B. and Schumm, S. A. (1958). Yield of sediment in relation to mean annual precipitation. *Am. Geophys. Union Trans.* **39**: 1076–1084.

Laronne, J. B. and Wilhelm, R. (2001). Shifting stage-volume curves: predicting event sedimentation rate based on reservoir stratigraphy. In D. J. Anthony, M. D. Harvey, J. B. Laronne, and M. P. Mosley, eds.,

Applying Geomorphology to Environmental Management. Littleton, CO: Water Resources Publication. p. 33–54.

Leckie, D. A. (1994). Canterbury Plains, New Zealand – implications for sequence stratigraphic models. *Am. Assoc. Petrol. Geol. Bull.* **78**: 1240–1256.

Leeder, M. R. and Alexander, J. (1987). The origin and tectonic significance of asymmetrical meander belts. *Sedimentology* **34**: 217–226.

Leliavsky, S. (1955). *An Introduction to Fluvial Hydraulics.* London: Constable.

Leopold, L. B. and Bull, W. B. (1979). Base level, aggradation, and grade. *Am. Philos. Soc. Proc.* **123**: 168–202.

Leopold, L. B. and Maddock, T., Jr. (1953). *The Hydraulic Geometry of Stream Channels and Some Physiographic Implications.* U.S. Geological Survey Professional Paper No. 252, pp. 1–57.

Leopold, L. B. and Wolman, M. G. (1957). *River Channel Patterns: Braided, Meandering and Straight.* U.S. Geological Survey Professional Paper No. 282-B, pp. 39–85.

LeStrange, G. (1905). *The Lands of the Eastern Caliphate.* London: Frank Cass.

Lewin, J. (1987). Historical river channel changes. In K. J. Gregory, J. Lewin and J. B. Thornes, eds., *Palaeohydrology in Practice.* Chichester: Wiley, pp. 161–175.

Lindley, E. S. (1919). Regime channels. *Punjab Enginr. Congr.* **7**: 63–74.

Livesey, R. H. (1963). *Channel Armoring Below Fort Randall Dam.* U.S. Department of Agriculture, Miscellaneous Publication No. 970, pp. 461–470.

Lofgren, B. E. (1969). Land subsidence due to the application of water. *Geol. Soc. Am. Rev. Eng. Geol.* **II**: 271–303.

Love, D. W. (1983). Summary of the late Cenozoic geomorphic and depositional history of Chaco Canyon. In S. G. Wells, D. N. Love and T. W. Gardner, eds., *Chaco Canyon Country.* American Geomorphology Field Group Guidebook, Albuquerque, NM: Adobe Press. pp. 187–194.

(1992). Rapid adjustment of the Rio Puerco to meander cutoff: implications for effective geomorphic processes, crossing thresholds and timing of events. *New Mexico Geological Society Guidebook, 43rd Field Conference,* San Juan Basin, No. IV, pp. 399–405.

Love, D. W. and Young, J. D. (1983). Progress report on the late Cenozoic geological evolution of the lower Rio Puerco. *New Mexico Geological Society Guidebook, 34th Field Conference, Socorro Region,* Albuquerque, NM: Adobe Press. pp. 277–284.

Lowe, M. (1993). *Debris-flow Hazards: A Guide for Land use Planning. Davis County, Utah.* U.S. Geological Survey Professional Paper No. 1519.

Ludwig, W. and Probst, J. (1996). *A Global Modeling of the Climatic, Morphological and Lithological Control of River Sediment Discharge to the Oceans.* International Association of Hydrologic Science Publication No. 236, pp. 21–28.

Mackay, A. L. (1991). *A Dictionary of Scientific Quotations.* Bristol: Adam Hilger, p. 297.

Mackay, D. (1945). Ancient river beds and dead cities. *Antiquity* **19**: 1935–1944.

Mackin, J. H. (1948). Concept of the graded river. *Geol. Soc. Am. Bull.* **59**: 463–512.

Macklin, M. G., Rumsby, B. T. and Heap, T. (1992). Flood alluviation and entrenchment: Holocene valley-floor development and transformation in the British uplands. *Geol. Soc. Am. Bull.* **104**: 631–643.

Maddock, T., Jr. (1960). Erosion control on Five Mile Creek, Wyoming. International Association of Scientific Hydrology Publication No. 53 (Helsinki), pp. 170–181.

Magilligan, F. (1992). Thresholds and the spatial variability of flood power during extreme floods. *Geomorphology* **5**: 373–390.

Marshall, J. (1931). *Mohen jo Daro and the Indus Civilization*. London: Arthur Probsthain.

Martinson, H. A. (1986). Channel adjustments after passage of a lahar. *Fourth Federal Interagency Sedimentation Conference Proceedings*, vol. 2, 5-143-5-152.

Mattes, M. J. (1969). The great Platte River road. *Nebraska State Hist. Soc. Pub.* **25**.

Mayer, L. (1985). Tectonic geomorphology of the Basin and Range Colorado Plateau boundary in Arizona. In M. Morisawa and J. T. Hack, eds., *Tectonic Geomorphology*. Boston: Allen and Unwin, pp. 235–259.

McAndrew, F. T. (1993). *Environmental Psychology*. Pacific Grove, CA: Brooks/Cole Publishing Co.

McCall, E. (1984). *Conquering the Rivers*. Baton Rouge, LA: Louisiana State University Press.

McCarthy, T. S., Ellery, W. N. and Stanistreet, I. G. (1992). Avulsion mechanisms on the Okavango fan, Botswana: the control of a fluvial system by vegetation. *Sedimentology* **39**: 779–796.

McDowell, P. F. (2001). Spatial variations in channel morphology at segment and reach scales, Middle Fork John Day River, northeastern Oregon. In J. M. Dorava, D. R. Montgomery, B. B. Palcsak and J. Fitzpatrick, eds., *Geomorphic Processes and Riverine Habitat*. American Geophysical Union Water Science and Application No. 4, pp. 159–172.

McRae, L. E. (1990). Paleomagnetic isochrons, unsteadiness, and non-uniformity of sedimentation in Miocene fluvial strata of the Siwalik Group, northern Pakistan. *J. Geol.* **98**: 423–428.

Meade, R. H. (1996). River-sediment inputs to major deltas. In J. D. Milliman and B. U. Haq, eds., *Sea-level Rise and Coastal Subsidence*. Dordrecht: Kluwer Academic Publisher, pp. 63–85.

Meade, R. H., Yuzyk, T. R., and Day, T. J. (1990). Movement and storage of sediment in rivers of the United States and Canada. In M. G. Wolman and H. C. Riggs, eds., *Surface Water Hydrology: The Geology of North America*, vol. 0–1. New York: Geological Society of America, pp. 255–280.

Melton, F. A. (1936). An empirical classification of floodplain streams: *Geogr. Rev.* **26**: 593–609.

Melton, F. A. (1959). Aerial photographs and structural geology. *J. Geol.* **67**: 351–370.

Memon, M. M. (1969). Alluvial morphology of the lower Indus plain and its relation to land use. *Pakistan Geogr. Rev.* **24**: 1–33.

Mercer, A. G. (1992). Impact of future water resources projects on the Nile. *Nile 2000, Conference on Protection and Development of the Nile and Other major rivers*, No. 1, 2-8-1–2-8-6.

Merritt, R. H. (1984). *The Corps, the Environment and the Upper Mississippi River basin*. Washington, D.C.: Historical Div., Office of Chief of Engineers.

Miall, A. D. (1996). *The Geology of Fluvial Deposits*. Berlin: Springer.

Millar, R. G. (2000). Influence of bank vegetation on alluvial channel patterns. *Water Resour. Res.* **36**: 1109–1118.

Miller, A. J. and Gupta, Avijit (Eds.) (1999). *Varieties of Fluvial Form*. Chichester: Wiley.

Miller, D. (1985). Ideology and the Harappan civilization. *J. Anthropol. Archaeol.* **4**: 34–71.

Milliman, J. D. and Meade, R. H. (1983). World-wide delivery of river sediment to the oceans. *J. Geol.* **91**: 1–21.

Milliman, J. D., Quraishee, G. S, and Beg, M. A. A. (1984). Sediment discharge from the Indus River to the ocean: Past, present and future. In B. V. Haq and J. D. Milliman, eds., *Marine Geology and Oceanography of Arabian Sea and Coastal Pakistan*. New York: Van Nostrand Reinhold, pp. 65–70.

Mollard, J. D. (1973). *Airphoto Interpretation of Fluvial Features: Fluvial Processes and Sedimentation*. Edmonton: Proceedings of Hydrology Symposium, University Alberta, pp. 341–380.

Molnar, P. (1987). The distribution of intensity associated with the great 1987 Assam earthquake and the bounds on the extent of the rupture zone. *J. Geol. Soc. (India.)* **30**: 13–27.

Montgomery, D. R., Abb, T. B., Buffington, J. M., Peterson, N. P., Schmidt, K. M. and Stock, J. D. (1996). Distribution of bedrock and alluvial channels in forested mountain drainage basins. *Nature* **381**: 587–589.

Montgomery, D. R. and Buffington, J. M. (1997). Channel-reach morphology in mountain drainage basins. *Geol. Soc. Am. Bull.* **109**: 596–611.

Morgan, J. P. and McIntire, W. G. (1959). Quaternary geology of the Bengal Basin, East Pakistan and India. *Geol. Soc. Am. Bull.* **70**: 319–342.

Morisawa, M. and LaFlure, E. (1979). Hydraulic geometry, stream equilibrium and urbanization. In D. D. Rhodes and G. P. William, eds., *Adjustments of the Fluvial System*. Dubuque: Kendall-Hunt, pp. 333–350.

Mosley, M. P. (1987). The classification and characterization of rivers. In K. Richards, ed., *River Channels*, Oxford: Blackwell, pp. 295–320.

Moss, J. H. and Kochel, R. C. (1978). Unexpected geomorphic effects of the Hurricane Agnes storm and flood, Conestoga drainage basin, Southeastern Pennsylvania. *J. Geol.* **86**: 1–11.

Mughal, M. R. (1982). Recent archaeological research in the Cholistan Deserts in Possehl, G. L. (ed.) *Harappan Civilization*. New Delhi: Oxford and IBH Publishing Co.

Mughal, M. R. (1990). The protohistoric settlement patterns in the Choliston Desert. *South Asian Archaeology, 9th International Conference of the Association of South Asian Archaeologists* (Venice), pp. 143–156.

Mulhern, P. F. (1975). Form and pattern on the Gros Ventre River. Unpublished MS thesis, Colorado State University.

Mycielska-Dowgiallo, E. (1977). Channel pattern changes during the last glaciation and Holocene in the northern part of the Sandomirez basin in the middle part of the Vistula valley, Poland. In K. G. Gregory, ed., *River Channel Changes*. New York: Wiley Interscience, pp. 75–87.

Nadler, C. T. and Schumm, S. A. (1981). Metamorphosis of South Platte and Arkansas Rivers, eastern Colorado. *Phys. Geogr.* **2**: 95–115.

Nanson, G. C. and Croke, J. C. (1992). A genetic classification of floodplains. *Geomorphology* **4**: 459–486.

Nanson, G. C. and Hean, D. S. (1985). The West Dapto flood of February 1984: rainfall characteristics of channel changes. *Aus. Geogr.* **16**: 249–257.

Nanson, G. C. and Huang, H. Q. (1999). Anabranching rivers: Divided efficiency leading to fluvial diversity. In A. J. Miller and A. Gupta, eds., *Varieties of Fluvial Form*. Chichester: Wiley, pp. 477–494.

Nanson, G. C. and Knighton, A. D. (1996). Anabranching rivers: their cause, character and classification. *Earth Surf. Process. Landf.* **21**: 217–239.

Nanson, G. C. and Young, R. W. (1981a). Downstream reduction of rural channel size with contributing urban effects in small coastal streams of Southeastern Australia. *J. Hydrol.* **52**: 239–255.

Nanson, G. C. and Young, R. W. (1981b). Overbank deposition and floodplain formation on small coastal streams of New South Wales. *Zeit. Geomorph.* **25**: 332–347.

Nelson, B. W. (1970). Hydrography, sediment dispersal and recent historical development of the Po River delta, Italy. In J. P. Morgan and R. H. Shaver, eds., *Deltaic Sedimentation, Modern and Ancient.* Tulsa: Special Publication of the Society of Economic Paleontologists and Mineralogists No. 15, pp. 52–184.

Nevins, T. H. F. (1965). River classification with particular reference to New Zealand. *Fourth New Zealand Geographical Conference Proceedings*, pp. 83–90.

Ning, Q. (1990). Fluvial processes in the lower Yellow River after levee breaching at Tongwaxiang in 1855. *J. Sediment. Res.* 5: 1–13.

Nissen, H. J. (1988). *The Early History of the Ancient Near East, 9,000–2,000 B.C.* Chicago: University Chicago Press. 215 p.

Odemerho, F. O. (1984). The effects of shifting cultivation on stream channel size and hydraulic geometry in small headwater basins of southwestern Nigeria. *Geograf. Ann., Ser. A.* **66**: 327–340.

Ohmori, H. (1982). *Functional Relationship between the Erosion Rate and the Relief Structure in the Japanese mountains.* University of Tokyo, Department of Geography, Bulletin No. 14, pp. 65–74.

(1983). *Erosion Rates and their Relation to Vegetation from the Viewpoint of World-wide Distribution.* University of Tokyo, Department of Geography, Bulletin No. 15, pp. 77–91.

(1992). Dynamics and erosion rate of the river running on a thick deposit supplied by a large landslide. *Zeit. Geomorph.* **36**: 129–140.

(1996). Morphological characteristics of longitudinal profiles of rivers in the South Island, New Zealand. University of Tokyo, Department of Geography. Bulletin No. 28, pp. 1–23.

Oldham, C. F. (1893). The Saraswati and the lost river to the Indian desert. *J. Royal Asiatic Soc.,* **34**: 49–76.

Oldham, R. D. (1926). The Cutch earthquake of 16 June 1819 with a revision of the great earthquake of 12 June 1897. *Geol. Surv. India Mem.* **46**: 1–77.

Osterkamp, W. R., Johnson, W. C., and Dixon, M. D. (2001). Biophysical gradients related to channel islands, middle Snake River, Idaho. In Dorava, J. M., Montgomery, D. R., Palcsak, B. B., and Fitzpatrick, F. A., eds., *Geomorphic Processes and Riverine Habitat.* American Geophysic Union, Water Science and Application No. 4, pp. 73–83.

Ouchi, S. (1985). Response of alluvial rivers to slow active tectonic movement. *Geol. Soc. Am. Bull.* **96**: 504–515.

Patton, P. C. and Schumm, S. A. (1965). Gully erosion, northwestern Colorado: A threshold phenomenon. *Geology* 3: 83–90.

Paulissen, E. and Vermeersch, P. M. (1987). Earth, man, and climate in the Egyptian Nile Valley during the Pleistocene. In Close, A. E. ed., *Prehistory of Arid North Africa.* Dallas: Southern Methodist University Press.

Peake, J., Peterson, M., and Lavstrump, M. (1985). Interpretation of vegetation encroachment and flow relationships in the Platte River by use of remote sensing techniques. Omaha: Remote-Sensing Applications Laboratory, Department of Geography – Geology, University of Nebraska at Omaha.

Penick, J., Jr., (1981). *The New Madrid Earthquakes of 1811–1812*. Columbia, MO: Univ. Missouri Press.

Peterson, M. S. (1986). *River Engineering*. Englewood Cliffs, NJ: Prentice-Hall.

Petts, G. E. (1979). Complex response of river channel morphology subsequent to reservoir construction. *Prog. Phys. Geogr.* **3**: 329–362.

Petts, G. E. (1995). Changing river channels: The geographic tradition. In A. Gurnell and G. Petts, eds., *Changing River Channels*. New York: Wiley, pp. 1–23.

Phillips, J. D. (2002). Human impacts on the environment: Unpredictability and the primacy of place. *Physical Geogr.* **22**: 321–332.

Pickup, G. (1986). Fluvial landforms. In D. N. Jeans, ed., *Australia: A geography. The Natural Environment*. Sydney: Sydney University Press, pp. 148–179.

Piégay, H. and Schumm, S. A. (2003). System approaches in fluvial geomorphology. In G. M. Kondolf and H. Piégay, eds., *Tools in fluvial geomorphology*. Chichester: Wiley, pp. 105–134.

Piggott, S. (1953). A forgotten empire of antiquity. *Sci. Am.* **189**: 42–48.

Popov, I. V. (1962). Application of morphological analysis to the evaluation of the general channel deformations of the River Ob. *Soviet Hydrol.* 267–324.

Potter, P. E. (1978). Significance and origin of big rivers. *J. Geol.* **86**: 13–33.

Possehl, G. L. (1967). The Mohen jo Daro Floods: A reply. *Am. Anthropol.* **69**: 32–40.

Postans, T. (1843). Memoranda on the Rivers Nile and Indus. *J. Royal Asiatic Soc.* **7**: 273–280.

Prokopovich, N. P. (1983). Neotectonic movement and subsidence caused by piezometric decline. *Assoc. Eng. Geol. Bull.* **20**: 393–404.

Proshansky, H. M. (1990). The pursuit of understanding: An intellectual history. In I. Altman and K. Christensen, eds., *Environment and Behavior Studies: Emergence of Intellectual Traditions*. New York: Plenum.

Prosser, I. P. and Slade, C. J. (1994). Gully formation and the role of valley-floor vegetation, southeastern Australia. *Geology* **22**: 1127–1130.

Rahn, P. H. (1977). Erosion below mainstem dams on the Missouri River, *Assoc. Engineering Geol.* **14**: 157–181.

Raikes, R. L. (1968). Kalibangan: death from natural causes. *Antiquity* **42**: 286–291.

Raikes, R. L. and Dales, G. F. (1977). The Mohen jo Daro floods reconsidered. *J. Palaeontol. Soc. India* **20**: 251–260.

Reid, J. B., Jr. (1992). The Owens River as a tiltmeter for Long Valley Caldera, California. *J. Geol.* **100**: 353–363.

Rennell, J. 1765. *A General Map of the River Baramputrey, from its Confluence with the Ifsamuty near Pacca towards Assam, India*. London: Office Library and Records Office.

Rice, S. and Church, M. (1998). Grain size along two gravel-bed rivers: Statistical variation, spatial pattern and sedimentary links. *Earth Surf. Process. Landf.* **23**: 345–363.

Richards, K. S. (1979). Channel adjustment to sediment pollution by the China Clay industry in Cornwall, England. In D. D. Rhodes and G. P. Williams, eds., *Adjustments of the Fluvial System*, Dubuque: Kendall-Hunt, pp. 309–332.

Richards, K., Chandra, S., and Friend, P. (1993). Avulsive channel systems, characteristics and examples. In J. L. Best and C. S. Bristow, eds., *Braided Rivers*, Bath: Special Publication of the Geological Society, London, No. 75, pp. 195–203.

Rodolfo, K. S. (1989). Origin and early evolution of lahar channel at Mabinit, Mayon Volcano, Philippines. *Geol. Soc. Am. Bull.* **101**: 414–426.

Rosgen, D. L. (1994). A classification of natural rivers, *Catena* **22**: 169–199.

Ross, C. P. (1923). *The Lower Gila Region, Arizona.* U.S. Geological Survey Water-Supply Paper No. 498.

Ross, D. R. (1982). *Style and Significance of Surface Deformation in the Vicinity of New Madrid, MO.* U.S. Geological Survey Professional Paper No. 1236-H, pp. 95–114.

Rubey, W. W. (1952). *Geology and Mineral Resources of the Hardin and Brussels Quadrangles (in Illinois).* U.S. Geological Survey Professional Paper No. 218.

Russell, R. J. (1954). Alluvial morphology of Anatolian rivers. *Ann. Assoc. Amer. Geogrs.* **44**: 363–391.

Rutherfurd, I. (1999). Some human impacts on Australian stream channel morphology. In S. Brizga and B. Finlayson, eds., *River Management – The Australian Experience*, Chichester: Wiley, pp. 11–49.

Rutherfurd, I. (2001). Storage and movement of slugs of sand in a large catchment: Developing a plan to rehabilitate the Glenelg River, SE Australia. In D. J. Anthony, M. D. Harvey, J. B. Laronne and M. P. Mosley, eds., *Applying Geomorphology to Environmental management.* Highlands Ranch, CO: Water Resources Publications, pp. 309–334.

Saggs, H. W. F. (1989). *Civilizations before Greece and Rome.* New Haven: Yale University Press.

Said, R. (1981). *The Geological Evolution of the River Nile.* New York: Springer-Verlag.

Saucier, R. T. (1981). Current thinking on riverine processes and geologic history as related to human settlement in the southeast. *Geosci. Man* **22**: 7–18.

(1994). *Geomorphology and Quaternary History of the Lower Mississippi Valley.* Vicksburg, MS: Mississippi River Commission.

Saucier, R. T. and Fleetwood, A. R. (1970). Origin and chronologic significance of late Quaternary terraces, Ouchita River, Arkansas and Louisiana. *Geol. Soc. Am. Bull.* **81**: 869–890.

Schama, S. (1995). *Landscape and Memory.* New York: Knopf.

Schattner, I. (1962). The lower Jordan River. *Hebrew Univ. Jerusalem. Publ.* **11**.

Schmidt, K. M. and Montgomery, D. R. (1985). Limits to relief. *Science* **270**: 617–620.

Schnitter, N. J. (1994). *A History of Dams.* Rotterdam: Balkema.

Schumann, R. R. (1989). Morphology of Red Creek, Wyoming, an arid-region anastomosing channel system. *Earth Surf. Process. Landf.* **14**: 277–288.

Schumm, S. A. (1956). Evolution of drainage systems and slopes in badlands at Perth Amboy, New Jersey. *Geol. Soc. Am. Bull.,* **67**: 597–696.

(1960). *The Shape of Alluvial Channels in Relation to Sediment Type.* U.S. Geological Survey Professional Paper No. 352-B.

(1961). *Effect of Sediment Characteristics on Erosion and Deposition in Ephemeral-Stream Channels.* U.S. Geological Survey Professional Paper No. 352-C, pp. 31–70.

(1963). Sinuosity of alluvial rivers on the Great Plains. *Geol. Soc. Am. Bull.* **74**: 1089–1100.

(1968). *River Adjustment to Altered Hydrologic Regimen – Murrumbidgie River and Paleo Channels, Australia.* U.S. Geological Survey Professional Paper No. 598.

(1977). *The Fluvial System.* New York: Wiley.

(1979). Geomorphic thresholds: the concept and its applications. *Inst. British Geogrs. Trans.* **4**: 485–515.

(1981). *Evolution and Response of the Fluvial System, Sedimentologic Implications.* Society of Economic Paleontologists and Mineralogists Special Publication No. 31, pp. 19–29.

(1991). *To Interpret the Earth.* Cambridge: Cambridge University Press.

(1993). River response to baselevel change: Implications for sequence stratigraphy. *J. Geol.* **101**: 279–294.

(1999). Causes and Controls of Channel Incision. In S. E. Darby and A. Simon, eds., *Incised River Channels: Processes, Forms Engineering and Management.* New York: Wiley pp. 19–33.

Schumm, S. A. and Beathard, R. M. (1976). Geomorphic thresholds: an approach for river management. In *Rivers No. 76.* New York: American Society of Civil Engineers, pp. 1655–1679.

Schumm, S. A. and Chorley, R. J. (1983). *Geomorphic Controls on the Management of Nuclear Waste.* U.S. Nuclear Regulatory Commission. Report No. NUREG/CR-3276.

Schumm, S. A. and Hadley, R. F. (1957). Arroyos and the semiarid cycle of erosion. *Am. J. Sci.* **225**: 161–174.

Schumm, S. A. and Khan, H. R. (1972). Experimental study of channel patterns. *Geol. Soc. Am. Bull.* **83**: 1755–1770.

Schumm, S. A. and Lichty, R. W. (1963). *Channel Widening and Flood-plain Construction along Cimarron River in Southwestern Kansas.* U.S. Geological Survey Professional Paper No. 352-D.

Schumm, S. A. and Phillips, W. (1986). Composite channels of the Canterbury Plain, New Zealand: a martian analog? *Geology* **14**: 326–329.

Schumm, S. A. and Thorne, C. R. (1989). Geologic and geomorphic controls on bank erosion. In M. A. Ports, ed., *Hydraulic Engineering,* New York: Proceedings of the American Society of Civil Engineers, pp. 106–111.

Schumm, S. A. and Winkley (Eds.) (1994). *The Variability of Large Alluvial Rivers.* New York: American Society of Civil Engineers Press.

Schumm, S. A., Boyd, K. F., Wolff, C. G., and Spitz, W. J. (1995). A groundwater sapping landscape in the Florida Panhandle. *Geomorphology* **12**: 281–297.

Schumm, S. A., Dumont, J. F., and Holbrook, J. M. (2000). *Active Tectonics and Alluvial Rivers.* Cambridge: Cambridge University Press.

Schumm, S. A., Erskine, W. D., and Tilleard, J. (1996). Morphology, hydrology and evolution of the anastomosing Ovens and King Rivers, Australia. *Geol. Soc. Am. Bull.* **108**. 1212–1224.

Schumm, S. A., Harvey, M. D., and Watson, C. C. (1984). *Incised Channels. Morphology, Dynamics and Control.* Littleton, CO: Water Resources Publications.

Schumm, S. A., Mosley, M. P., and Weaver, W. E. (1987). *Experimental Fluvial Geomorphology.* New York: Wiley.

Schumm, S. A., Rutherfurd, I. D., and Brooks, J. (1994). Pre-cutoff morphology of the lower Mississippi River. In Schumm, S. A. and Winkley, B. R., eds., *The Variability of Large Alluvial Rivers,* New York: American Society of Civil Engineers Press, pp. 13–44.

Schuster, R. L. and Costa, J. E. (1986). A perspective on landslide dams. In R. L. Schuster, ed., *Landslide Dams.* Geotechnical Special Publication No. 3. New York: American Society for Civil Engineers, pp. 1–20.

Sedell, J. R. and Froggatt, J. L. (1984). Importance of streamside forests to large rivers: The isolation of the Willamette River, Oregon, USA from its floodplain by snagging and streamside forest removal. *Verhand. Internatl. Verein. Limnol.* **22**: 1828–1834.

Shalash, S. (1980). The effects of the High Aswan Dam on the hydrological regime of the River Nile. *Proceedings of Helsinki Symposium*, IAHS Publication No. 130, June.

Shalash, S. (1983). Degradation of the River Nile. *Internat. Water Power Dam Construc.* **35**(8): 56–58.

Shallat, T. (1994). *Structures in the Stream*. Austin, TX: University of Texas Press.

Shepherd, R. G. (1979). River channel and sediment responses to bedrock lithology and stream capture, Sandy Creek drainage, central Texas. In D. D. Rhodes and G. P. Williams, eds., *Adjustments of the Fluvial System*. Dubuque: Kendall-Hunt, pp. 255–276.

Shimizu, O. and Araya, T. (2001). Saru River. In T. Marutani, G. J. Brierley, N. A. Trustum and M. Page, eds., *Source-to-Sink Sedimentary cascades in Pacific rim geo-systems*. Motomachi: Matsumoto Sabo Work Office, Ministry of Land, Infrastructure and Transport, pp. 62–67.

Shroba, R. R., Schmidt, P. E., Crosby, E. J., and Hansen, W. R. (1979). *Storm and flood of July 31-August 1, 1976, in the basins, Larimer and Weld Counties, Colorado, Big Thompson River and Cache la Poudre River*. U.S. Geological Survey Professional Paper No. 1115, pp. 87–152.

Shroder, J. F., Jr. (1989). Hazards of the Himalaya. *Am. Sci.* **77**: pp. 564–573.
(1994). *Gradation Processes and Channel Evolution in Modified West Tennessee Streams: Process, Response and Form*. U.S. Geological Survey Professional Paper No. 1470.

Simons, D. B., Schumm, S. A., Chen, Y. H., and Beathard, R. M. (1976). A geomorphic study of Pool 4 and tributaries of the upper Mississippi River. Unpublished Report for U.S. Fish and Wildlife Service, Dept. of Civil Eng., Colorado State Univ., Fort Collins, CO.

Simons, R. K. and Simons, D. B. (1994). An analysis of Platte River channel changes. In S. A. Schumm and B. R. Winkley, eds., *The Variability of Large Alluvial Rivers*. New York: American Society of Civil Engineers Press, pp. 341–361.

Simpson, C. J. and Smith, D. G. (2001). The braided Milk River, northern Montana fails the Leopold-Wolman discharge-gradient test. *Geomorphology* **41**: 337–353.

Smith, D. G. (1976). Effect of vegetation on lateral migration of anastomosed channels of a glacier meltwater river. *Geol. Soc. Am. Bull.* **87**: 857–860.

Smith, D. G. (1979). Effects of channel enlargement by river ice processes on bankfull discharge in Alberta, Canada. *Water Resour. Res.* **15**: 469–475.

Smith, D. G. (1980). River ice processes. In D. R. Coates and J. D. Vitek, eds., *Thresholds in Geomorphology*. London: Allen & Unwin, pp. 323–344.

Smith, N. D. and Smith, D. G. (1984). William River: an outstanding example of channel widening and braiding caused by bed-load addition. *Geology* **12**: 78–82.

Smith, N. D., Cross, T. A., Dufficy, J. P., and Clough, S. R. (1989). Anatomy of an avulsion. *Sedimentology* **36**: 1–23.

Snelgrove, A. K. (1979). Migrations of the Indus River, Pakistan in response to plate tectonic motions. *J. Geol. Soc. India* **20**: 392–403.

Spitz, W. J. and Schumm, S. A. (1997). Tectonic geomorphology of the Mississippi valley between Osceola, Arkansas and Friars Point, Mississippi. *Engin. Geol.* **46**: 259–280.

Starkel, L. (1983). The reflection of hydrologic changes in the fluvial environment of the temperate zone during the last 15,000 years. In K. J. Gregory, ed., *Background to paleohydrology*. New York: Wiley, pp. 213–235.

Stevens, M. A. (1994). The Citanduy, Indonesia – one tough river. In S. A. Schumm and B. R. Winkley, eds., *The Variability of Large Alluvial Rivers*. New York: American Society of Civil Engineers Press, pp. 201–219.

Stevens, M. A., Simons, D. B., and Schumm, S. A. (1975a). Man-induced changes of middle Mississippi River. *J. Waterways, Harb. Coast. Eng.* **101**: pp. 119–133.

Stevens, M. A., Simons, D. B., and Richardson, E. V. (1975b). Non-equilibrium river form. *J. Hydraulics* **101**: 557–566.

Stein, A. (1942). A survey of the ancient sites along the "lost" Sarasvati River. *Geogr. J.* **99**: 173–182.

Summerfield, M. A. and Hulton, N. J. (1994). Natural controls of fluvial denudation rates in major world drainage basins. *J. Geophys. Res.* **99**: 13871–13883.

Thomas, M. S. and Anderson, J. B. (1989). Glacial eustatic controls on seismic sequences and para-sequences of the Trinity/Sabine incised valley, Texas continental shelf. *Gulf Coast Assoc. Geol. Soc. Trans.* **39**: 563–569.

Thorne, C. R. (1997). Channel types and morphological classification. In C. R. Thorne, R. D. Hey and M. D. Newson, eds., *Applied Fluvial Geomorphology for River Engineering and Management*. Chichester: Wiley, pp. 175–222.

Thorne, C. R. and Baghirathan, V. R. (1994). Blueprint for morphologic studies. In S. A. Schumm and B. R. Winkley, eds., *The Variability of Large Alluvial Rivers*. New York: American Society of Civil Engineers Press, pp. 441–454.

Thorne, C. R., Hey, R. D., and Newson, M. D. (Eds.) (1997). *Applied fluvial geomorphology for river engineering and management*. Chichester: Wiley.

Thornthwaite, C. W., Sharpe, C. F. S., and Dosch, E. F. (1942). *Climate and Accelerated Erosion in the Arid and Semi-arid southwest, with special Reference to the Polacca Wash Drainage Basin, Arizona*. U.S. Deptartment of Agriculture Technical Bulletin No. 808.

Tinkler, K. J. and Wohl, E. E. (Eds.) (1998). *Rivers over Rock: Fluvial Processes in Bedrock Channels*. Washington, D.C.: American Geophysical Union.

Todd, O. J. and Eliassen, S. (1940). The Yellow River problem. *Trans. Am. Soc. Civ. Eng.* **105**: 346–416.

Tooth, S. (2000). Downstream changes in dryland river channels: the Northern Plains of arid central Australia. *Geomorphology* **34**: 33–54.

Tooth, S. and Nanson, G. C. (2000). The role of vegetation in the formation of anabranching channels in an ephemeral river, Northern plains, arid central Australia. *Hydrol. Process.* **14**: 3099–3117.

Trimble, S. W. (1974). *Man-induced Soil Erosion on the Southern Piedmont 1700–1970*, Ankeny, IA: Soil Conservation Society of America.

Trimble, S. W. (1988). The impact of organisms on overall erosion rates within catchments in temperate regions. In H. Viles, ed., *Biogeomorphology*. Oxford: Basil Blackwell, pp. 183–242.

Trimble, S. W. (1997). Stream channel erosion and change resulting from riparian forests. *Geology* **25**: 467–469.

Trimble, S. W. and Mendel, A. C. (1995). The cow as a geomorphic agent a critical review. *Geomorphology* **13**: 233–253.

Vandaele, Poesen, J., Govers, G., and Van Wesemael, B. (1996). Geomorphic threshold conditions for ephemeral gully incision. *Geomorphology* **16**: 161–173.

Vandenberghe, J. (1993). The role of rivers in paleoclimatic reconstruction. *Paleoklimaforschung* **14**: 11–19.

Vandenberghe, J. (2001). A typology of Pleistocene cold-based rivers. *Quatern. Internl.* **79**: 111–121.

Van Gelder, A., Van Den Berg J. H., Cheng, and G. Xue, C. (1994). Overbank and channelfill deposits of the modern Yellow River delta. *Sediment. Geol.* **90**: 293–305.

Veatch, A. C. (1906). *Geology of the Underground Water Resources of Northern Louisiana and Southern Arkansas.* U.S. Geological Survey Professional Paper No. 46.

Walker, R. G. (1990). Perspective-facies modeling and sequence stratigraphy. *J. Sediment. Petrol.* **60**: 777–786.

Walters, W. H., Jr. (1975). Regime changes of the lower Mississippi River. Unpublished M.S. thesis, Colorado State University.

Walters, W. H., Jr. and Simons, D. B. (1984). Long-term changes of lower Mississippi River meander geometry. In Elliott, C. M. (ed.) *River Meandering.* New York: American Society of Civil Engineers, pp. 318–29.

Ware, E. F. (1911). *The Indian War of 1864.* Lincoln NE: Bison Books, University of Nebraska Press. (Reprinted 1963.).

Warner, R. F. (1983). Channel changes in the sandstone and shale reaches of the Nepean River, New South Wales. In R. W. Young and G. C. Nanson, eds., *Aspects of Australian Sandstone Landscapes.* New Zealand Geomorphology Group, Special Publication No. 1, pp. 106–119.

(1987a). Spatial adjustments to temporal variations in flood regime in some Australian rivers. In K. Richards, ed., *River Channels: Environment and Process.* Oxford: Blackwell, pp. 14–40.

(1987b). *The Impacts of Alternating Flood and Drought-dominated Regimes on Channel Morphology at Penrith, New South Wales, Australia.* IAHS Publication No. 168, pp. 327–338.

Webb, R. H., Melis, P. G., and Griffiths, P. G. *et al.* (1999). *Lava Falls Rapid in Grand Canyon: Effects of Late Holocene Debris Flows on the Colorado River.* U.S. Geological Survey Professional Paper No. 1591.

Wendland, W. M. and Bryson, R. A. (1974). Dating climatic episodes in Holocene. *Quart. Res.* **4**: 9–24.

Wendorf, F., Said, R., and Schild, R. (1970). Egyptian prehistory: Some new concepts. *Science* **169**: 1161–1171.

Wenke, R. J. (1990). *Patterns in Prehistory.* Oxford: Oxford University Press.

Wheeler, M. (1966). *Civilizations of the Indus Valley and Beyond.* New York: McGraw-Hill.

(1968). *The Indus Civilization.* Cambridge: Cambridge University Press.

Whipple, K. X., Snyder, N. P., and Dollenmayer, K. (2000). Rates and processes of bedrock incision by the Upper Ukak River since the 1912 Novarupta ash flow in the valley of Ten Thousand Smokes, Alaska. *Geology* **28**: 835–838.

Wilhelmy, H. (1969). Das Vrstromtal am Ostrand der Indusebene und das Sarasvati-problem. *Zeit Geomorphol.* **8** (Suppl): 76–93.

Williams, G. P. and Wolman, M. G. (1984). *Downstream Effects of Dams on Alluvial Rivers*. U.S. Geological Survey Professional Paper No. 1286.

Wilson, K. V. (1979). *Changes in Channel Characteristics, 1938–1974 of the Homochitto River and Tributaries, Mississippi*. U.S. Geological Survey Open-File Report No. 79–554.

Winkley, B. R. (1977). *Man-made Cutoffs on the Lower Mississippi River: Conception, Construction, and River Response*. U.S. Army Engineer District, Vicksburg, Corps of Engineers, Potamology Investigations, Report No. 300–2.

(1994). Response to the lower Mississippi River to flood control and navigation improvements. In S. A. Schumm and B. R. Winkley, eds., *The Variability of Large Alluvial Rivers*. New York: American Society of Civil Engineers, pp. 45–74.

Wittfogel, K. (1957). *Oriental Despotism*. New Haven: Yale University Press.

Wohl, E. E. (2000a). Anthropogenic impacts on flood hazards. In E. E. Wohl, ed., *Inland Flood Geomorphology*. Cambridge: Cambridge University Press. pp. 104–141.

(2000b). Geomorphic effects of floods. In E. E. Wohl, ed., *Inland Flood Geomorphology*. Cambridge: Cambridge University Press, pp. 167–193.

(2001). *Virtual Rivers*. New Haven: Yale University Press.

Wohl, E. E. and Ikeda, H. (1998). Pattern of bedrock channel erosion on the Bosco Peninsula, Japan. *J. Geol.* **106**: 331–345.

Wolman, M. G. and Eiler, J. P. (1958). Reconnaissance study of erosion and deposition by the flood of August 1955 in Connecticut. *Am. Geophys. Union* Trans. **39**: 1–14.

Wood, L., Etheridge, F. G., and Schumm, S. A. (1993). *The Effects of Rate of Base-level Fluctuation on Coastal Plain, Shelf and Slope Depositional Systems: An Experimental Approach*. International Association of Sedimentology, Special Publication No. 18, pp. 43–53.

Xu, J. (1996). Underlying gravel layers in a large sand-bed river and their influence on downstream-dam channel adjustment. *Geomorphology* **17**: 351–359.

Xu, Q., Wu, C., Yang, X., and Zhang, N. (1996). Paleochannels on the North China Plain: Relationships between their development and tectonics. *Geomorphology* **18**: 27–35.

Yodis, E. G. and Kesel, R. H. (1993). The effects and implications of base-level changes to Mississippi River tributaries. *Zeit. Geomorph.* **37**: 385–402.

Yoxall, W. H. (1969). The relationship between falling baselevel and lateral erosion in experimental streams. *Geol. Soc. Am.* **80**: 1379–1384.

Zhou, Z. and Pan, X. (1994). Lower Yellow River. In S. A. Schumm and B. R. Winkley, eds., *The Variability of Large Alluvial Rivers*. New York: American Society of Civil Engineers Press, pp. 363–94.

Index

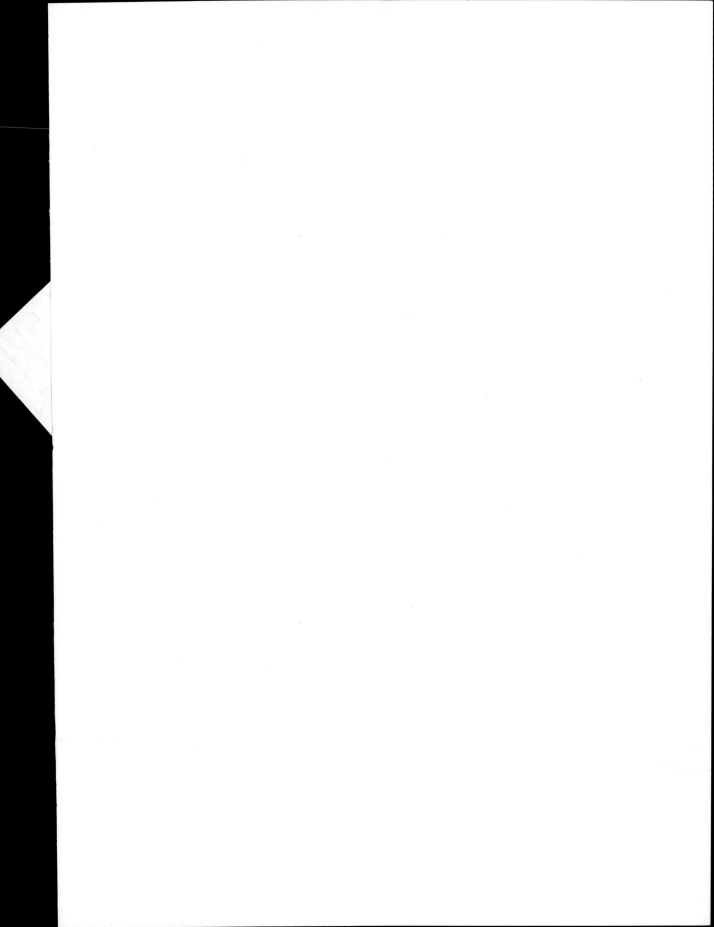

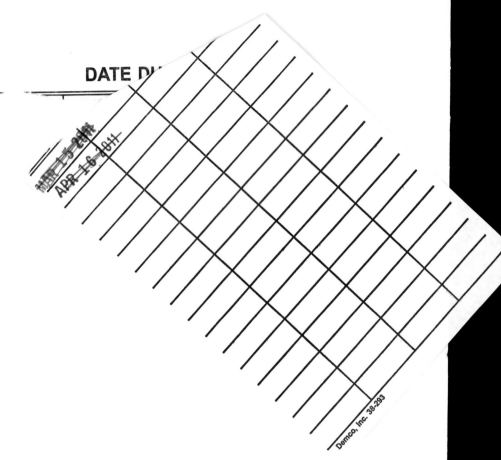

DATE DUE

Demco, Inc. 38-293

DATE DU